FEDERICO GARCIA LORCA ELEMENTARY
3231 N. Springfield Ave.
Chicago, IL 60618
(773) 534-0950

Everyday Mathematics®

The University of Chicago School Mathematics Project

STUDENT REFERENCE BOOK

Mc
Graw
Hill
Education

The University of Chicago School Mathematics Project

Max Bell, Director, *Everyday Mathematics* First Edition
James McBride, Director, *Everyday Mathematics* Second Edition
Andy Isaacs, Director, *Everyday Mathematics* Third, CCSS, and Fourth Editions
Amy Dillard, Associate Director, *Everyday Mathematics* Third Edition
Rachel Malpass McCall, Associate Director, *Everyday Mathematics* CCSS and Fourth Editions
Mary Ellen Dairyko, Associate Director, *Everyday Mathematics* Fourth Edition

Authors
Max Bell
Jean Bell
John Bretzlauf
Amy Dillard
James Flanders
Robert Hartfield
Andy Isaacs
Catherine Randall Kelso
James McBride
Kathleen Pitvorec
Peter Saecker

Writers
Lisa J. Bernstein
Andy Carter
Jeanne Di Domenico
Lila K.S. Goldstein
Jesch Reyes
Elizabet Spaepen
Judith S. Zawojewski

Digital Development Team
Carla Agard-Strickland, Leader
John Benson
Gregory Berns-Leone
Scott Steketee

Technical Art
Diana Barrie, Senior Artist
Cherry Inthalangsy

UCSMP Editorial
Elizabeth Olin
Kristen Pasmore
Molly Potnick

Contributors
Lance Campbell
Rosalie A. DeFino
Kathryn Flores
Gina Garza-Kling
Rebecca W. Maxcy

www.everydaymath.com

Copyright © McGraw-Hill Education

Send all inquiries to:
McGraw-Hill Education
8787 Orion Place
Columbus, OH 43240

ISBN: 978-0-02-143697-2
MHID: 0-02-143697-5

Printed in the United States of America.

1 2 3 4 5 6 7 8 9 DOW 20 19 18 17 16 15

Contents

About the *Student Reference Book*. ix

How to Use the *Student Reference Book* x

Standards for Mathematical Practice　　1

Mathematical Practices 2

Problem Solving: Make Sense and Keep Trying. 4

Create and Make Sense of Representations. 8

Make Sense of Conjectures and Arguments 10

Create and Use Mathematical Models 12

Choose Tools to Solve Problems 14

Be Precise and Accurate 16

Look for Structure in Mathematics 20

Create and Justify Rules and Shortcuts. 22

Guide to Solving Number Stories. 26

A Problem-Solving Diagram. 27

Problem Solving and the Mathematical Practices 28

Operations and Algebraic Thinking　　31

Algebra 32

Relations 33

Parentheses 34

Order of Operations 36

Using Variables 37

Multiplication Properties 39

Multiplication Fact Strategies 42

Extended Facts 45

Number Models and Situation Diagrams 47

Estimation 52

Arrays and Factors of Counting Numbers. 53

Kinds of Counting Numbers 54

Multiples 55

Multiplicative Comparisons. 56

Contents

Number Patterns 58

Frames-and-Arrows Diagrams 62

Function Machines and "What's My Rule?" Problems 65

Sound, Music, and Mathematics 69

Number and Operations in Base Ten 75

Uses of Numbers 76

Kinds of Numbers 77

Place Value for Whole Numbers 78

Expanded Form 80

Comparing Numbers and Amounts 81

Estimation 82

Front-End Estimation 84

Rounding 85

Close-But-Easier Numbers 88

Addition Methods 90

Subtraction Methods 94

Extended Multiplication Facts 102

Area Models for Multiplication 103

Multiplication Methods 106

Basic Division Facts 109

Extended Division Facts 110

Division Methods 111

Machines that Calculate 117

Number and Operations—Fractions 123

Fractions 124

Reading and Writing Fractions 125

Fractions Equal To and Greater than One 127

Meanings of Fractions 128

Uses of Fractions 129

Fraction Circles 130

Using Fraction Strips 132

Using Fractions to Name Points on a Number Line 133

Equivalent Fractions. 136

Finding Equivalent Fractions 137

Number-Lines Poster 138

Equivalent Fractions on a Ruler. 139

An Equivalent Fraction Rule 141

Table of Equivalent Fractions 142

Renaming Fractions Equal To or Greater Than One 143

Renaming Mixed Numbers 144

Comparing Fractions 145

Introducing Decimals 149

Understanding Decimals 150

Extending Place Value to Decimals 152

Comparing Decimals 154

Problem Solving by Drawing and Reasoning about
Fractions 156

Estimating with Fractions 158

Adding and Subtracting Fractions with Like
Denominators 160

Adding Mixed Numbers with Like Denominators 162

Subtracting Mixed Numbers with Like Denominators 164

Adding Tenths and Hundredths 166

Finding Fractions of a Set. 170

Fractions as Multiples of Unit Fractions 171

Multiplying Fractions by Whole Numbers 173

Multiplying Mixed Numbers by Whole Numbers 175

Measurement and Data   177

Natural Measures and Standard Units 178

The Metric System and the U.S. Customary System 179

Length: Metric System. 180

Contents

Length: U.S. Customary System 184

Mass 188

Weight 190

Liquid Volume: Metric System 193

Liquid Volume: U.S. Customary System 195

Time 198

Perimeter 200

Area 202

Area of a Rectangle 204

Standard Units for Measuring Angles 207

Measuring and Drawing Angles 208

Finding Unknown Angle Measures 211

Tally Charts and Bar Graphs 213

Line Plots 214

Making Sense of Data 216

Measurements in the Natural World 217

Geometry — 223

Geometry in Our World 224

Points and Line Segments 226

Rays and Lines 227

Angles 228

Classifying Angles 229

Parallel Lines and Segments 230

Line Segments, Rays, Lines, and Angles 231

Polygons 232

Triangles 233

Quadrilaterals 234

The Geometry Template 236

Line Symmetry 238

Mathematics and Architecture 239

Games — 245

Games 246

Angle Add-Up 248

Angle Race 249

Angle Tangle 250

Beat the Calculator (Extended Facts) 251

Buzz Games 252

Decimal Top-It 253

Divide and Conquer 254

Division Arrays 255

Division Dash 256

Factor Bingo 257

Factor Captor 258

Fishing for Digits 259

Fishing for Fractions 260

Fraction/Decimal Concentration 262

Fraction Match 263

Fraction Multiplication Top-It 264

Fraction Top-It 265

How Much More? 266

Multiplication Wrestling 267

Name That Number 268

Number Top-It 269

Polygon Capture 270

Product Pile-Up 271

Rugs and Fences 272

Spin-and-Round 273

Subtraction Target Practice 274

Top-It Games 275

Contents

Real-World Data 277

Introduction 278
Length of Roller Coasters. 279
Sizes of Indoor Water Parks 280
Sizes of Zoos around the World 281
Major U.S. City Populations in 2010 282
Major U.S. City Populations in 1930 283
Normal September Rainfall (in centimeters) 284
Normal Monthly Precipitation (in centimeters) 285
What Do Americans Eat? 286
Food Supplies around the World 287
Songbird Wing Lengths 288
Data Sources 289

Tables and Charts 291

Appendix 294

Glossary 301

Answer Key 319

Index 328

About the *Student Reference Book*

The *Student Reference Book* is a helpful guide to review math concepts and skills, a resource to find the meaning of math terms, and interesting reading when you want to learn about a new topic in mathematics.

A reference book is organized to help readers find information quickly and easily. Dictionaries, encyclopedias, atlases, and cookbooks are examples of reference books. Reference books are not like novels and biographies, which you often read in order from beginning to end. When you read reference books, you look for specific information at the time you need it. Then you read just the pages you need at that time.

You can use this *Student Reference Book* to look up and review information on topics in mathematics. It includes the following information:

- A **table of contents** that lists the topics and gives an overview of how the book is organized

- Essays describing how to use **mathematical practices** to solve problems and show mathematical thinking

- Essays on **mathematical content,** such as algebraic thinking, numbers, operations, fractions, measurement, data, and geometry

- A collection of tables and maps that includes **real-world data**

- A collection of **photo essays** that show in words and pictures some of the ways that scientists, artists, engineers, and others have used mathematics throughout history or how they use it today

- Directions on how to play **mathematical games** to practice your math skills

- A set of **tables** and **charts** that summarize information, such as a place-value chart, rules for the order of operations, and tables of equivalent values

- An **appendix** that includes directions on how to use a **calculator** to perform various mathematical operations

- A **glossary** of mathematical terms consisting of definitions and some illustrations

- An **answer key** for the **Check Your Understanding** problems

- An **index** to help you locate topics quickly

- **Videos** and **interactive problems** available through the electronic version of this book in the Student Learning Center

How to Use the *Student Reference Book*

As you work in class or at home, you can use the *Student Reference Book* to help you solve problems. For example, when you don't remember the meaning of a word or aren't sure what method to use, you can use the *Student Reference Book* as a tool.

You can look in the **table of contents** or the **index** to find pages that give a brief explanation of the topic. The explanation will often include definitions of important math words and show examples of problems with step-by-step sample solutions.

While reading the text, you can take notes that include words, pictures, and diagrams to help you understand what you are reading. Work through the examples and try to follow each step.

At the end of some of the essays, you will find problems in **Check Your Understanding** boxes. Solve these problems and then check the **answer key** at the back of the book. These exercises will help you make sure that you understand the information you have been reading. Make sense of the problems by comparing your answers with those in the answer key. If necessary, work backward from the sample answers to revise your work.

The **Standards for Mathematical Practice** section includes interesting problems that you can solve. The discussions illustrate how fourth-grade students use the practices to solve these problems.

The world of mathematics is a very interesting and exciting place. Read the **photo essays,** explore **real-world data,** or review topics learned in class. The *Student Reference Book* is a great place to continue your investigation of math topics and ideas.

Once you are familiar with the overall structure of the *Student Reference Book*, you can use it to read about different mathematical concepts. As you follow your interests, you will find that your skills as an independent reader and problem-solver will improve.

Standards for Mathematical Practice

Mathematical Practices

People who use mathematics develop ways of working that help them solve problems. These ways of doing mathematics are called **mathematical practices.**

Look at each picture below.

In what ways might these students be using mathematical practices?

The next page lists the Standards for Mathematical Practice (SMPs) you will use in *Everyday Mathematics*. Below each standard is a list of Goals for Mathematical Practice (GMPs) that can help you understand what it means to use the practices.

In this section, you will see how some fourth-grade students use mathematical practices as they solve problems and reason about mathematics. As you read, think about the ways you can develop these practices to become a more powerful problem solver.

Mathematical Practice 1: Make sense of problems and persevere in solving them.

GMP1.1 Make sense of your problem.

GMP1.2 Reflect on your thinking as you solve your problem.

GMP1.3 Keep trying when your problem is hard.

GMP1.4 Check whether your answer makes sense.

GMP1.5 Solve problems in more than one way.

GMP1.6 Compare the strategies you and others use.

Mathematical Practice 2: Reason abstractly and quantitatively.

GMP2.1 Create mathematical representations using numbers, words, pictures, symbols, gestures, tables, graphs, and concrete objects.

GMP2.2 Make sense of the representations you and others use.

GMP2.3 Make connections between representations.

Mathematical Practice 3: Construct viable arguments and critique the reasoning of others.

GMP3.1 Make mathematical conjectures and arguments.

GMP3.2 Make sense of others' mathematical thinking.

Mathematical Practice 4: Model with mathematics.

GMP4.1 Model real-world situations using graphs, drawings, tables, symbols, numbers, diagrams, and other representations.

GMP4.2 Use mathematical models to solve problems and answer questions.

Mathematical Practice 5: Use appropriate tools strategically.

GMP5.1 Choose appropriate tools.

GMP5.2 Use tools effectively and make sense of your results.

Mathematical Practice 6: Attend to precision.

GMP6.1 Explain your mathematical thinking clearly and precisely.

GMP6.2 Use an appropriate level of precision for your problem.

GMP6.3 Use clear labels, units, and mathematical language.

GMP6.4 Think about accuracy and efficiency when you count, measure, and calculate.

Mathematical Practice 7: Look for and make use of structure.

GMP7.1 Look for mathematical structures such as categories, patterns, and properties.

GMP7.2 Use structures to solve problems and answer questions.

Mathematical Practice 8: Look for and express regularity in repeated reasoning.

GMP8.1 Create and justify rules, shortcuts, and generalizations.

Problem Solving: Make Sense and Keep Trying

Fourth graders are solving a special kind of number puzzle called a Magic Track.

Start with any number greater than or equal to 100. Put it in one of the boxes. Add or subtract the numbers as you move around the track. See how the numbers change as you go around.

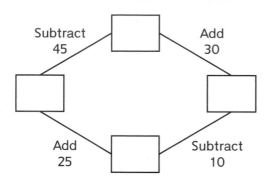

Try the Magic Track yourself. What happens?

Ms. Martin asks students to try using the Magic Track.

Rachel tries 150 and starts in the bottom box. First she goes around the track to the right. She adds and subtracts this way:

$150 - 10 = 140$ $140 + 30 = 170$ $170 - 45 = 125$ $125 + 25 = 150$

When she gets back to the bottom box, she has 150 again. Then she tries adding and subtracting going around the track to the left. She still gets 150 when she gets back to the bottom box.

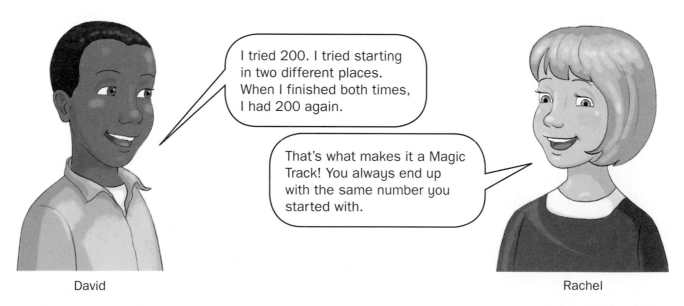

I tried 200. I tried starting in two different places. When I finished both times, I had 200 again.

That's what makes it a Magic Track! You always end up with the same number you started with.

David

Rachel

GMP1.1 Make sense of your problem.

When students tried the Magic Track with their own numbers and in their own ways, they were making sense of the puzzle.

Ms. Martin asks students to make their own Magic Tracks.

Jada shows her Magic Track to the class.

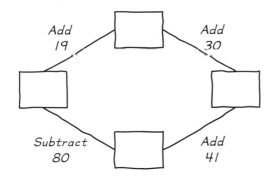

Try making your own Magic Track.

Jada's classmates try her Magic Track to see what happens.

David starts with 100 and ends at 110, so he thinks that the Magic Track doesn't work.

Rachel puts in 150 and goes around the track both ways. Both times she ends at 160.

Mia starts at the bottom and puts in 200. She gets 210, so she agrees that this Magic Track doesn't work.

Ms. Martin asks for ways to fix Jada's broken Magic Track.

I noticed that everyone's final answer is off by 10. Maybe we can fix the Magic Track by subtracting 70 instead of 80.

I tried Ethan's idea and started with 120. But I ended up with 140. That's even further off. I think we need to subtract more instead of less. Let's subtract 90 instead.

Ethan

Jada

GMP1.2 Reflect on your thinking as you solve your problem.
Jada realized that subtracting 70 made the ending number even further from the starting number, so she decided to subtract 90 instead. She was reflecting on her thinking as she solved the problem.

Jada shows how to fix her broken Magic Track.

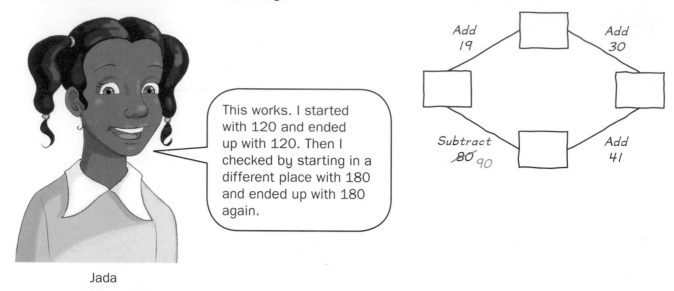

This works. I started with 120 and ended up with 120. Then I checked by starting in a different place with 180 and ended up with 180 again.

Jada

GMP1.3 Keep trying when your problem is hard.
When Jada saw that her first Magic Track didn't work, she kept trying and figured out a way to fix it instead of giving up.

Ms. Martin asks if there are any other ways to fix the broken Magic Track.

Mia fixed the broken Magic Track this way:

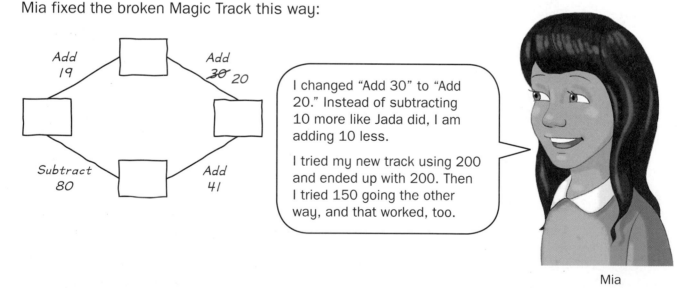

I changed "Add 30" to "Add 20." Instead of subtracting 10 more like Jada did, I am adding 10 less.

I tried my new track using 200 and ended up with 200. Then I tried 150 going the other way, and that worked, too.

Mia

GMP1.4 Check whether your answer makes sense.
Mia checked to see if her new Magic Track worked by trying 200 and 150 in two different ways.

GMP1.6 Compare the strategies you and others use.
When Mia explained that subtracting 10 more is like adding 10 less, she was comparing her strategy to Jada's.

Ms. Martin asks students to think of a way to predict whether a Magic Track will work before trying a number in it.

Jada replies, "I think a Magic Track works when the amounts added balance out the amounts subtracted. My broken Magic Track didn't work because my added numbers totaled 90 but my subtracted number was just 80. I fixed it by subtracting 90 so that the numbers would balance."

Jada writes on the whiteboard:

Broken Magic Track

Added: $19 + 30 + 41 = 90$
Subtracted: 80

Fixed Magic Track

Added: $19 + 30 + 41 = 90$
Subtracted: 90

Mia tells the class, "When I fixed Jada's Magic Track, I found a different way to balance the numbers. I made the amount added and the amount subtracted both equal 80 instead of 90. As long as you add the same amount that you subtract, you'll end up with the same number that you started with."

Mia writes on the whiteboard:

Mia's fixed Magic Track

Added: $19 + 20 + 41 = 80$
Subtracted: 80

GMP1.5 Solve problems in more than one way.
Jada fixed her Magic Track by changing the amount subtracted to 90 to match the amount added. Mia fixed the Magic Track by changing the amount added to 80 to match the amount subtracted. They solved the problem in more than one way.

Mathematical Practice 1: Make sense of problems and persevere in solving them.

Create and Make Sense of Representations

Some fourth-grade students are solving a multiplication problem. They use mathematical **representations** such as numbers, words, pictures, symbols, or real objects to show their thinking.

Here is the problem: $6 * 32 = ?$

Jasmine uses repeated addition to represent $6 * 32$.

I know that $6 * 32$ means six equal groups of 32. So I added 32 six times and got 192.

Jasmine

$$
\begin{array}{r}
1 \\
3\ 2 \\
3\ 2 \\
3\ 2 \\
3\ 2 \\
3\ 2 \\
+\quad 3\ 2 \\
\hline
1\ 9\ 2
\end{array}
$$

Ethan knows that multiplication can be represented by using the area of a rectangle. He draws a rectangle to represent $6 * 32$.

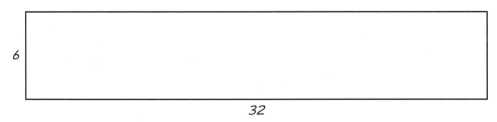

Then Ethan partitioned the rectangle into smaller rectangles. He split the length of the rectangle into numbers that are easy to multiply mentally. He added the areas of the smaller rectangles to find the answer of 192.

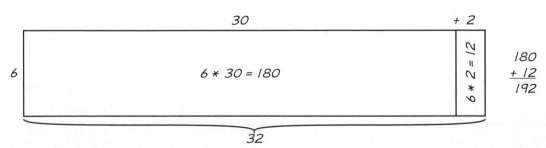

GMP2.1 Create mathematical representations using numbers, words, pictures, symbols, gestures, tables, graphs, and concrete objects.

Jasmine used a number sentence showing repeated addition for her representation. Ethan created a representation using an area diagram and number sentences.

David uses partial products to multiply.

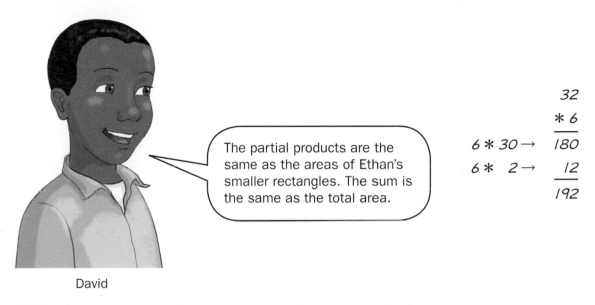

The partial products are the same as the areas of Ethan's smaller rectangles. The sum is the same as the total area.

$$
\begin{array}{r}
32 \\
* 6 \\
\hline
\end{array}
$$

$6 * 30 \rightarrow \quad 180$

$6 * \ \ 2 \rightarrow \quad \underline{\ 12}$

$\quad\quad\quad\quad\quad 192$

David

GMP2.2 Make sense of the representations you and others use.

GMP2.3 Make connections between representations.

David created a number sentence by using partial products and made a connection to Ethan's representation. When he noticed that his partial products and answer were the same as in Ethan's area representation, David was making sense of his own representation.

When you reason about a problem using different representations, you are thinking abstractly. When you think about numbers and amounts, you are thinking quantitatively.

Mathematical Practice 2: Reason abstractly and quantitatively.

Check Your Understanding

How is Ethan's representation like the one below? How is it different?

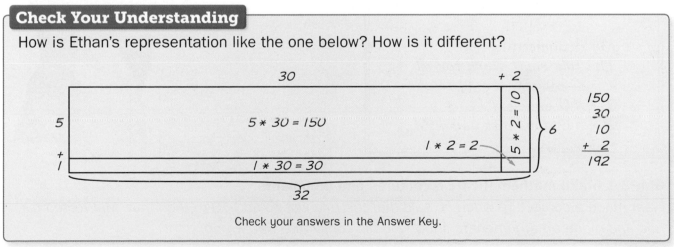

Check your answers in the Answer Key.

Make Sense of Conjectures and Arguments

Some fourth graders are making conjectures and arguments as they solve problems. A **conjecture** is a statement that might be true. In mathematics, conjectures are not simply guesses. They are based on information and mathematical thinking. Mathematical reasoning that shows whether a conjecture is true or false is called an **argument.** Mathematical arguments can use words, pictures, or symbols.

Ms. Martin asks her students to solve this "What's My Rule?" problem.

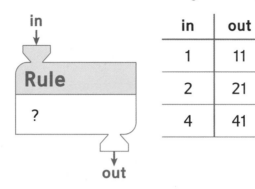

in	out
1	11
2	21
4	41

Liam writes his conjecture on the board.

Liam

My Conjecture:
The rule is to add 10 to the input.

Mia makes an argument about Liam's conjecture.

Mia

My Argument:
The rule must work for all three rows:
1 + 10 equals 11
2 + 10 does not equal 21

Any rule must work for all three rows. Liam's rule works for the first row, but it doesn't work for the second row.

Mia

GMP3.1 Make mathematical conjectures and arguments.

Liam made a conjecture when he said that the rule was to add 10 to the input. Mia tested his rule and made an argument to show why his conjecture was wrong.

Now I see why my rule doesn't work. I have a new rule, and I can explain why it works. I'll write my conjecture and argument on the board.

Liam

Liam

My New Conjecture:
The rule is to multiply
the input by 10,
then add 1.

My Argument:
The rule works for all
three rows:
$(1 * 10) + 1 = 11$
$(2 * 10) + 1 = 21$
$(4 * 10) + 1 = 41$

It works! Your new rule works for all three rows.

Mia

GMP3.2 Make sense of others' mathematical thinking.

After Liam made sense of Mia's argument about his first conjecture, he revised his conjecture to be true for this problem.

Mathematical Practice 3: Construct viable arguments and critique the reasoning of others.

Create and Use Mathematical Models

You can create mathematical **models** to represent real-world situations using graphs, drawings, tables, symbols, number models, diagrams, or words. When you use a model to answer a question, think about the answer to see whether it makes sense in the real world. If not, revise the model to better fit the problem.

Fourth graders are solving this problem:

Ethan has a new scooter. Ethan, Tony, Jada, and Rachel each take turns riding it on the way to the playground that is 6 blocks away.

If the 4 friends share the scooter ride equally, what is the total distance that each student will ride on the scooter on the way to the playground?

Solve the problem. Does your model give you an answer that makes sense?

Ethan shows the class his model, and then Tony models the problem another way.

Block 1	Block 2	Block 3	Block 4	Block 5	Block 6
Ethan	Tony	Jada	Rachel	Ethan / Tony	Jada / Rachel

Each friend rides 1 block + $\frac{1}{2}$ block.

I drew 6 rectangles to represent the 6 blocks. I gave 1 block to each friend and had 2 blocks left over. Then I split each of the remaining blocks in half.

Each friend rides the scooter 1 block and then later $\frac{1}{2}$ of a block.

I also drew a rectangle to represent each of the blocks, but I split each of the blocks into fourths. I gave each friend $\frac{1}{4}$ of each block.

So each friend rides the scooter $\frac{1}{4}$ of each block.

Ethan

Tony

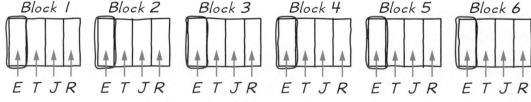

Block 1   Block 2   Block 3   Block 4   Block 5   Block 6
E T J R   E T J R   E T J R   E T J R   E T J R   E T J R

Each friend rides $\frac{1}{4} + \frac{1}{4} + \frac{1}{4} + \frac{1}{4} + \frac{1}{4} + \frac{1}{4}$ blocks.

GMP4.1 Model real-world situations using graphs, drawings, tables, symbols, numbers, diagrams, and other representations.

Ethan and Tony modeled the situation by drawing 6 rectangles to represent the 6 blocks to get to the playground. Ethan distributed 1 block to each friend and split the remaining 2 blocks among the friends. Tony split each of the blocks into equal portions for each friend.

Jada uses Ethan's and Tony's ideas to model the problem. She is thinking about distance, so she uses a number line and draws 4 hops that each show a distance of $1\frac{1}{2}$ blocks.

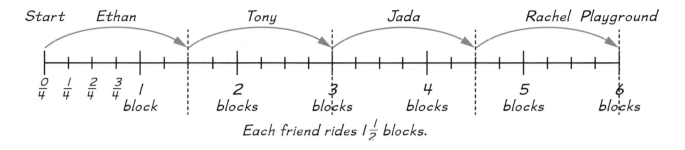

Jada

GMP4.2 Use mathematical models to solve problems and answer questions.

When Ethan and Tony modeled the problem in different ways, they both found ways to share the ride fairly. Jada used their ideas to model each friend taking one turn for $1\frac{1}{2}$ blocks, rather than more turns of smaller distances.

Mathematical Practice 4: Model with mathematics.

Choose Tools to Solve Problems

You can use many types of tools to solve problems in mathematics. Some examples of mathematical tools are pencil and paper, rulers, protractors, base-10 blocks, fraction circle pieces, diagrams, tables, charts, graphs, and calculators.

Think about this problem:

Mia and David ran on a track. Mia ran $\frac{6}{8}$ of the track's total distance and David ran $\frac{3}{4}$ of it. Who ran farther?

Liam uses fraction strips to solve the problem.

> How would you begin to solve this problem?
> What tools might you use?

> I chose the eighths strip so I could see $\frac{6}{8}$ and the fourths strip to show $\frac{3}{4}$. I lined them up so I could easily compare the fractions. $\frac{6}{8}$ lines up with $\frac{3}{4}$.

Liam

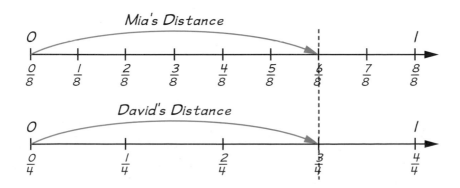

Jasmine draws number lines to show how far each student ran.

You're right, Liam. I used number lines, and $\frac{6}{8}$ and $\frac{3}{4}$ are each the same distance from 0. So, Mia and David ran the same distance.

Jasmine

GMP5.1 Choose appropriate tools.

Liam used fraction strips and Jasmine used number lines to show the part of the track that Mia and David ran.

Tony uses fraction circle pieces to solve the problem.

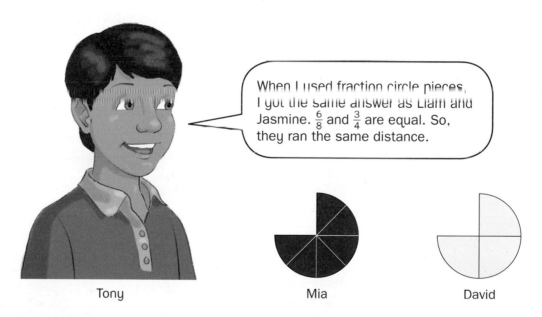

When I used fraction circle pieces, I got the same answer as Liam and Jasmine. $\frac{6}{8}$ and $\frac{3}{4}$ are equal. So, they ran the same distance.

Tony   Mia   David

GMP5.2 Use tools effectively and make sense of your results.

All three students used tools to represent the distance Mia and David ran on the track. They used the tools to show that $\frac{3}{4}$ and $\frac{6}{8}$ are equal.

Mathematical Practice 5: Use appropriate tools strategically.

Be Precise and Accurate

It is important to be **precise** and **accurate** when measuring, calculating, or counting. It is also important to use clear and precise language to explain your mathematical thinking and results.

A group of fourth graders are decorating a bulletin board. They want to cover the board with fabric and use paper trim to make a border around the edges. Ms. Martin gives them measurements for the fabric and paper trim and asks them to figure out if they have enough of each for the bulletin board.

Length of Bulletin Board

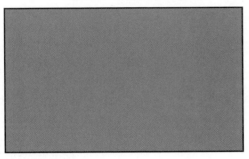

Width of Bulletin Board

Fabric: 2 yards long and 1 yard wide

Rachel quickly measures the bulletin board. She draws and labels a picture:

a little less than 2 yards

1 yard

1 yard

a little less than 2 yards

Roll of paper trim: 200 inches long

I used a yardstick to measure the length and the width. The width is 1 yard, and the length is a little less than 2 yards.

The width of the fabric is the same as the width of the bulletin board, 1 yard.

And since the length of the fabric, 2 yards, is longer than the bulletin board, we have more than enough fabric to cover the board.

Rachel

Tony

Diana Barrie

David

Now we have to figure out if we have enough paper trim to go around the edge of the bulletin board. I added the lengths of all four sides to find the perimeter of the board.

a little less than 2 yards

1 yard

1 yard

a little less than 2 yards

Perimeter is about 2 yd + 1 yd + 2 yd + 1 yd.
Perimeter is about 6 yards.

Rachel remembers that they have 200 inches of paper trim to go around the board. In order to compare the 200 *inches* of trim to the 6 *yards* around the board, she uses a calculator to figure out how many inches are equal to 6 yards.

1 yard = 36 inches
6 yards = 6 * 36 inches = 216 inches

Tony

The perimeter of the board is about 6 yards, so that's about 216 inches. We only have 200 inches of paper trim, so I guess we don't have enough.

Wait! I just estimated the length of the board. We need a closer measurement to know for sure. Instead of measuring to the nearest yard, we should measure the length of the board to the nearest inch.

Rachel

GMP6.2 Use an appropriate level of precision for your problem.
Rachel and Tony used appropriate levels of precision when they made decisions about how precise they needed to be. Tony knew he could use an estimate to figure out that they had enough fabric because it was clear that they had more than enough. However, Rachel realized that they needed a closer, or more precise, measurement than the estimate to know whether they had enough paper trim.

SRB

Tony starts to measure the bulletin board again.

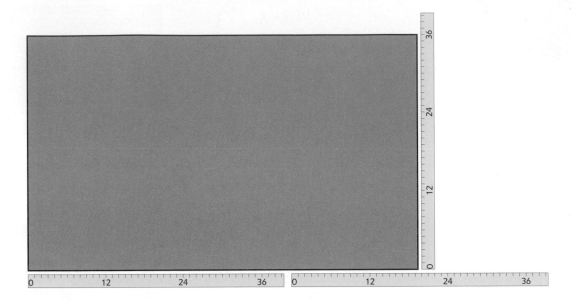

David reminds Tony that there shouldn't be any space between where the first yard ends and where the second yard begins.

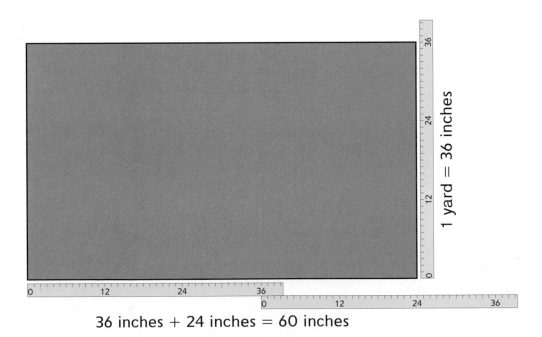

36 inches + 24 inches = 60 inches

Tony measures again. His new measurement shows that the length of the bulletin board is 36 inches for the first yard plus 24 inches of the second yard. The board is 60 inches long.

GMP6.4 Think about accuracy and efficiency when you count, measure, and calculate.

David was thinking about accuracy when he corrected Tony's measurement. He knew that there shouldn't be any space between the yards in order to get an accurate measurement.

Tony says, "We have enough trim." He shows Rachel a number model for the length of the paper trim that they need:

$$120 + 36 + 36 = 192$$

Rachel doesn't understand Tony's number model, so Tony rewrites it this way:

Perimeter = 60 inches + 60 inches + 36 inches + 36 inches = 192 inches

I needed to find the perimeter, or the distance around the bulletin board. First I added the lengths of the longer sides in my head. 60 inches + 60 inches = 120 inches. Then I added that 120 inches to the lengths of the shorter sides, 36 inches plus 36 inches, and figured out that the total distance around is 192 inches.

Tony

That makes more sense. And since 192 is less than 200, our 200 inches of paper trim is more than enough to go around the board.

Rachel

GMP6.1 Explain your mathematical thinking clearly and precisely.

After Rachel said she didn't understand Tony's number sentence, Tony explained his thinking clearly and precisely. He told her that he added the lengths of the sides to find the perimeter, and then he explained the steps he did in his head.

GMP6.3 Use clear labels, units, and mathematical language.

When Tony explained his thinking to Rachel, he used clear mathematical language. He labeled the measurements in his number sentence with inches as the unit.

Mathematical Practice 6: Attend to precision.

Look for Structure in Mathematics

An important part of doing mathematics is looking for **mathematical structures** such as **properties, categories,** and **patterns.** Finding these structures can help you solve problems and think about math in new ways.

Jada, Liam, and Ethan are looking at this dot pattern.

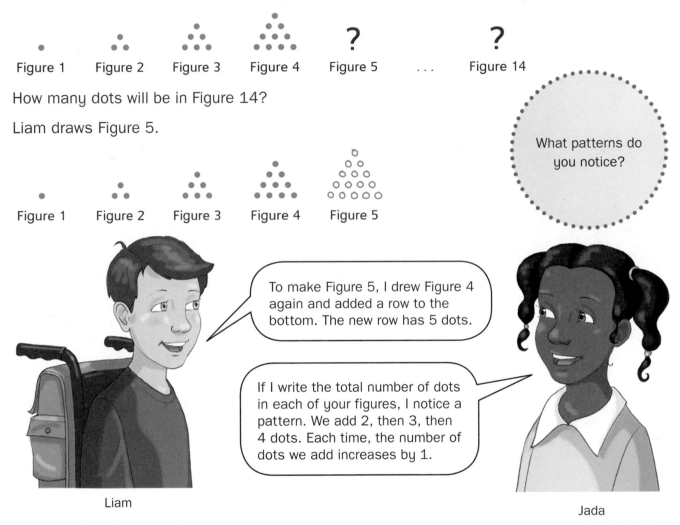

How many dots will be in Figure 14?

Liam draws Figure 5.

> To make Figure 5, I drew Figure 4 again and added a row to the bottom. The new row has 5 dots.

Liam

> If I write the total number of dots in each of your figures, I notice a pattern. We add 2, then 3, then 4 dots. Each time, the number of dots we add increases by 1.

> What patterns do you notice?

Jada

Jada writes the total number of dots above each figure to show the addition pattern she saw.

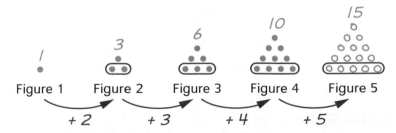

GMP7.1 Look for mathematical structures such as categories, patterns, and properties.
Jada was looking for structures when she noticed an addition pattern she could use to find the total number of dots in each triangle.

Ethan uses the pattern Jada noticed to make a table.

Figure	1	2	3	4	5	6	7	8	9	10	11	12	13	14
# of dots	1	3	6	10	15	21	28	36	45	55	66	78	91	105

+2 +3 +4 +5 +6 +7 +8 +9 +10 +11 +12 +13 +14

I used my table to extend the pattern. Figure 14 will have 105 dots. That's a lot of dots!

I thought about Figure 14 a different way. The bottom row of Figure 2 has 2 dots, and the bottom row of Figure 3 has 3 dots. If I continue that pattern, the bottom row of Figure 14 will have 14 dots.

Liam

Ethan

1 dot
⋮
11 dots
12 dots
13 dots
14 dots

Figure 14

$14 + 13 + 12 + 11 + 10 + 9 + 8 + 7 + 6 + 5 + 4 + 3 + 2 + 1 = 105$

Liam starts drawing a picture of Figure 14, then realizes that he doesn't need to finish the picture to know how many dots he will draw. He writes a number sentence to show the total number of dots in Figure 14 by adding the number of dots from each row. He adds using a calculator and finds the same answer as Ethan. Figure 14 will have 105 dots.

GMP7.2 Use structures to solve problems and answer questions.

Ethan used the pattern in his table to find the number of dots in Figure 14. Liam found the same answer when he used his pattern to write an addition number sentence.

Mathematical Practice 7: Look for and make use of structure.

Create and Justify Rules and Shortcuts

You can make your math work easier and faster if you create and justify rules and shortcuts that work for many problems.

Fourth graders are studying the pattern of dots below. For each figure, more dots are added to the pattern.

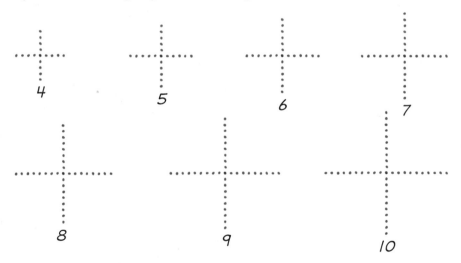

Figure 1   Figure 2   Figure 3   . . .   Figure 10

How would you figure out the number of dots in Figure 10?

Ms. Martin asks the class to figure out the number of dots in Figure 10.

Jasmine draws a picture showing Figures 4 through 10.

4   5   6   7

8   9   10

Ms. Martin asks other students to explain Jasmine's drawing.

I think Jasmine noticed that the number of dots in each arm going out from the middle goes up by one in each figure.

I think Jasmine drew the number of dots in each arm to be the same as the figure number.

Mia   Tony

Jasmine explains her drawing.

> Mia is right about how I drew the first couple of figures. But then I noticed the same thing as Tony. I drew the arms so that the number of dots in each arm matched the figure number.

Jasmine

Ms. Martin asks volunteers to report the number of dots in each arm and the total number of dots in Figure 10.

Mia explains, "I drew the same picture as Jasmine. I had 10 dots in each arm in Figure 10, and when I counted all the dots, I got 41 dots."

Tony adds to Jasmine's drawing in green. He explains his thinking, "I got 41 dots, too. But I figured out a shorter way. I used a number model to add the number of dots in each arm plus the red dot in the middle."

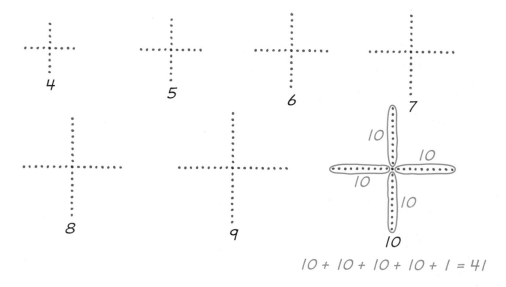

$$10 + 10 + 10 + 10 + 1 = 41$$

Ms. Martin asks Tony if he can use the same strategy for Figure 25. He writes a number model on the board and explains, "I don't have to draw the picture because I know the number of dots in each arm will be 25 dots, the same as the figure number. I added 25 four times because there are 4 arms. The 1 in the number model is for the middle dot."

$$25 + 25 + 25 + 25 + 1 = 101 \text{ dots}$$

Ms. Martin asks the students to try using Tony's strategy on Figure 15. She invites them to the board to show their work.

Mia	Jasmine	Tony
15 dots		
15 dots	(4 * 15) + 1	15 + 15 + 15 + 15 + 1
15 dots	60 + 1	30 + 30 + 1
15 dots		60 + 1
+1 dot	61 dots	61 dots
61 dots		

Mia explains why the different number sentences all show the same correct answer.

Everyone showed 15 dots for each of the four arms. Tony and I *added* 15 four times, but Jasmine did 4 *times* 15, and that means the same thing. The 1 in all the number sentences is for the dot in the center of the figure.

Mia

Ms. Martin asks the students to use their work to create a rule for finding the total number of dots in any figure in this dot pattern.

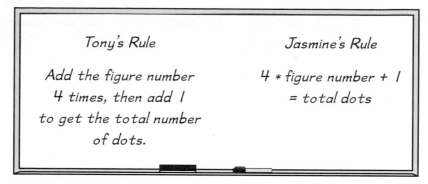

Tony's Rule

Add the figure number 4 times, then add 1 to get the total number of dots.

Jasmine's Rule

*4 * figure number + 1 = total dots*

GMP8.1 Create and justify rules, shortcuts, and generalizations.

Tony created a shortcut for finding the total number of dots without drawing a picture. He justified the strategy by connecting the parts of his number sentence to the four arms and the center dot in Figure 25. Tony and Jasmine were generalizing when they looked back at their work to create rules. Their rules work to find the total number of dots in any figure in the pattern.

Mathematical Practice 8: Look for and express regularity in repeated reasoning.

Check Your Understanding

1. Use Tony's or Jasmine's rule to find the number of dots in Figure 50.

2. Find the figure number for a figure with 401 dots. Show or tell how you know.

3. Write a rule for finding the figure number if you are given the total number of dots.

Check your answers in the Answer Key.

Guide to Solving Number Stories

Solving number stories is a big part of mathematics. Good problem solvers often follow a few simple steps every time they solve a number story. When you are solving a number story, you can follow these steps in any order.

Make sense of the problem.

- Read the problem. What do you understand?

- What do you know?

- What do you need to find out?

Make a plan.

What will you do?

- Add?
- Subtract?
- Multiply?
- Divide?
- Draw a picture?
- Draw a diagram?

- Make a table?
- Make a graph?
- Write a number model?
- Use counters or base-10 blocks?
- Make tallies?
- Use a number grid or number line?

Solve the problem.

- Show your work.

- Keep trying if you get stuck.

  → Reread the problem.

  → Think about the strategies you have already tried and try new strategies.

- Write your answer with the units.

Check.

Does your answer make sense? How do you know?

A Problem-Solving Diagram

The diagram below can help you think about problem solving. The boxes show the kinds of things you do when you use mathematical practices to solve problems. The arrows show that you don't always do things in the same order.

Organize the information.
- Study the information in the problem.
- Arrange the information into a list, table, graph, or diagram.
- Look for more information if you need it.
- Get rid of information you don't need.

Understand the problem.
- Retell the problem in your own words.
- Figure out what you want to find.
- Figure out what you know.
- Imagine what the answer will look like.
- Make a guess at the answer.

Play with the information.
- Draw a picture, diagram, or another mathematical representation.
- Write a number model.
- Model the problem using objects such as counters or base-10 blocks.

Check your answer as you work.
- Does your answer make sense?
- Compare your answer with a classmate's.
- Does your answer fit the problem?
- Can you solve the problem another way?

Figure out what math can help.
- Can you use addition? Subtraction? Another operation?
- Can you use geometry? Patterns? Other mathematics?
- Try the math. See what happens.
- What units are you using? Label your numbers with units.

Problem Solving and the Mathematical Practices

When you solve mathematical problems, you use mathematical practices. Ms. Martin asks her class to solve this problem.

Jacob's father is buying erasers as party favors for Jacob and his 9 guests.

The erasers are sold in boxes of 6 erasers. Each box has these 6 erasers:

Solve the problem. Think about the mathematical practices you use.

Jacob's father wants to give the same number of erasers to each of the 10 children at the party. He wants to buy the smallest possible number of boxes, and he doesn't want to have any erasers left over.

How many boxes of erasers should he buy?
How many erasers will each child get?

After students work on the problem, Ms. Martin asks volunteers to explain or show their solutions to the class.

David shows this number model to the class.

$$10 * 6 = 60$$

Jacob's father should buy 60 boxes. There are 10 kids at the party and there are 6 erasers in a box. That's 10 * 6.

I don't understand how you got 60 boxes. That would be a lot of erasers!

Whoops! I mean 60 erasers. 10 boxes * 6 erasers in each box is 60 erasers.

So that would be 10 boxes. One for each child.

Jasmine

David

Diana Barrie

Ms. Martin asks if anyone found a smaller number of boxes that would work. Mia shares her answer.

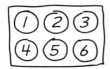 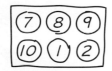 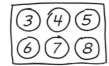 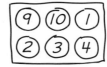 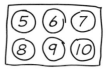

Jacob's dad only needs to buy 5 boxes of erasers. I drew rectangles with 6 circles in them to represent a box of erasers. I put numbers in each circle to represent each kid. I kept drawing boxes and numbering erasers until the 10th kid got the last eraser in a box. With 5 boxes, the number of kids and erasers came out even, so there are no erasers left over.

Mia

Ethan also shares his work.

	1 box	6 erasers	4 children don't get erasers.
	2 boxes	12 erasers	1 eraser each, 2 erasers left over.
	3 boxes	18 erasers	1 eraser each, 8 erasers left over.
	4 boxes	24 erasers	2 erasers each, 4 left over.
	5 boxes	30 erasers	3 erasers each, none left over!

I agree with Mia. 5 boxes have 30 erasers. So for 10 children, each gets 3 erasers. And when Jacob's dad passes out 3 erasers to each child, there are none left.

Ethan

Ms. Martin asks if anyone found an even smaller number of boxes that works.

You cannot have a smaller number of boxes. Look at Ethan's work. The first line says 1 box isn't enough. That's right, because there are only 6 erasers. And 2 boxes, 3 boxes, and 4 boxes all have leftovers. With 5 boxes, there are no leftover erasers.

So, 5 boxes has to be the smallest number of boxes that Jacob's father can buy.

Rachel

Check Your Understanding

Use the mathematical practices described on page 3, at the beginning of this section, to answer the questions.

1. What mathematical practices did David use?

2. What mathematical practices did Mia use?

3. What mathematical practices did Ethan use?

Check your answers in the Answer Key.

Operations and Algebraic Thinking

Algebra

Algebra is a part of mathematics that uses mathematical statements to describe patterns, model real-world situations, and show how numbers compare to one another. You use algebra when you write number models to represent a number story or real-world situations. Algebra is like arithmetic, but algebra uses letters, blanks, question marks, and other symbols to stand for amounts that are not known. Using these mathematical statements can help you solve problems more easily.

Mathematical Symbols

- **Digits** are 0, 1, 2, 3, 4, 5, 6, 7, 8, and 9.

- **Relation symbols** are $=$, $\neq$, $<$, and $>$.

- **Operation symbols** are $+$, $-$, $\times$ or $*$, and $\div$ or $/$.

Number Sentences

Number sentences are like English sentences, except that they use math symbols instead of words. Using symbols instead of words makes the sentences easier to write and work with.

You have seen number sentences that contain **digits, operations symbols,** and **relation symbols.** A number sentence *must* contain numbers and a relation symbol. It does not have to contain any operation symbols. Here are some number sentences:

$$7 = 35 / 5 \qquad 8 + 4 > 8 + 3 \qquad 8 * 8 \neq 63 \qquad 1{,}234 < 4{,}321$$

A number sentence may be **true** or **false.** For example, the number sentence $14 + 3 = 17$ is true, while the number sentence $12 = 9$ is false.

Open Sentences

In some number sentences, one or more numbers may be missing or unknown. In place of each missing number there is a letter, a blank, a question mark, or some other symbol. These number sentences are called **open sentences.** A symbol that stands for a missing or unknown number is called a **variable.** For example, $9 + x = 15$ is an open sentence in which x stands for some number. For most open sentences, you can't tell whether the sentence is true or false until you know which number replaces the variable.

- If you replace x with 10 in the open sentence $9 + x = 15$, you get the number sentence $9 + 10 = 15$, which is false.

- If you replace x with 6 in the open sentence $9 + x = 15$, you get the number sentence $9 + 6 = 15$, which is true.

The number 6 is a **solution** of the open sentence $9 + x = 15$, because $9 + 6 = 15$ is a true number sentence. Finding a solution for an open number sentence is called **solving the number sentence.**

Relations

A **relation** tells how two things compare. The table at the right shows common relations that compare numbers and their symbols.

Symbol	Relation
=	is equal to
≠	is not equal to
<	is less than
>	is greater than

Equations

A number has many different names. For example, 3 * 4 is equal to 12, and 2 * 6 is also equal to 12. So, 12 and 3 * 4 and 2 * 6 are just different ways to express the same number. We say that 12 and 3 * 4 and 2 * 6 are **equivalent** names because they all name the same number.

Everyday Mathematics uses **name-collection boxes** as a way to show equality.

- The label at the top of the box shows a number.

- The names written inside the box are equivalent names for the number on the label. Each of these names is equal to 12.

Another way to state that two things are equal is to write a number sentence using the = symbol. Number sentences containing the = symbol are called **equations.**

12
2 * 6
12 − 0
$\frac{12}{1}$
$\frac{1}{2}$ * 24
36 ÷ 3

a name-collection box for 12

Examples

Here are some equations. The third equation is false.

$$4 * 5 = 20 \qquad 12 - 4 = 8 + 0 \qquad 100 / 5 = 25 \qquad 5 = 5$$

Inequalities

Number sentences that do not contain an = symbol are called **inequalities.** Common relations in inequalities are > (is greater than), < (is less than), and ≠ (is not equal to).

Examples

Here are some inequalities. The second inequality is false.

$$5 + 6 < 15 \qquad 25 > 30 \qquad 36 \neq 7 * 6 \qquad 3 * 5 > 3 * 4$$

Check Your Understanding

Compare. Use =, <, or > to make each number sentence true.

1. 100 ☐ 55 + 45   **2.** $\frac{1}{2}$ ☐ 0.5   **3.** $\frac{3}{4}$ ☐ $\frac{1}{4}$

4. 3 * 50 ☐ 4 * 50   **5.** $\frac{1}{2}$ * 100 ☐ 50   **6.** 4.3 ☐ 4.25

Check your answers in the Answer Key.

Parentheses

Parentheses ( ) are grouping symbols that help you understand number sentences with more than one operation. Parentheses make the meaning of a number sentence clear.

Parentheses show which operation to do first.

Examples

Solve. $(15 - 3) * 2 = $ **?**

The parentheses tell you to subtract first.	$15 - 3 = 12$
Then multiply 12 by 2.	$12 * 2 = 24$
The answer is 24.	24

$(15 - 3) * 2 = $ **24**

If the same numbers are grouped differently, you may get a different answer.

Solve. $15 - (3 * 2) = $ **?**

The parentheses tell you to multiply first.	$3 * 2 = 6$
Then subtract 6 from 15.	$15 - 6 = 9$
The answer is 9.	9

$15 - (3 * 2) = $ **9**

Example

Solve. $9 - (3 - 2) = $ **_n_**

The parentheses tell you to solve $3 - 2$ first.	$9 - (3 - 2) = n$
Then subtract $9 - 1$.	$9 - 1 = n$
The answer is 8.	$8 = n$

$9 - (3 - 2) = $ **8**

Examples

$(2 + 3) * (4 + 5) = $ **?**

There are 2 sets of parentheses. Do the operations inside the parentheses first.	$(2 + 3) * (4 + 5) = $ **?**
Then multiply those answers.	$5 \quad * \quad 9 \quad = ?$
The final answer is 45.	$5 \quad * \quad 9 \quad = 45$

$(2 + 3) * (4 + 5) = $ **45**

Sometimes a number sentence does not have parentheses, and you are asked to include parentheses to make the number sentence true.

Example

Make this number sentence true by including parentheses:
$4 = 6 - 3 - 1$.

There are two possible places for the parentheses:

$? = (6 - 3) - 1$	or $? = 6 - (3 - 1)$
Subtract in the parentheses first.	Subtract in the parentheses first.
$6 - 3 = 3$	$3 - 1 = 2$
Then subtract again.	Then subtract again.
$3 - 1 = 2$	$6 - 2 = 4$
2 is not equal to 4.	4 is equal to 4.
The parentheses do *not* belong around $6 - 3$.	The parentheses do belong around $3 - 1$.

The correct place for the parentheses is $4 = 6 - (3 - 1)$.

Example

Include parentheses to make this statement true: $14 - 6 \div 2 = 4$.

There are two possible places for the parentheses:

$(14 - 6) \div 2 = ?$	or $14 - (6 \div 2) = ?$
Subtract first. $14 - 6 = 8$	Divide first. $6 \div 2 = 3$
Then divide. $8 \div 2 = 4$	Then subtract. $14 - 3 = 11$
4 is equal to 4.	11 is not equal to 4.
The parentheses do belong around $14 - 6$.	The parentheses do *not* belong around $6 \div 2$.

The correct place for the parentheses is $(14 - 6) \div 2 = 4$.

Check Your Understanding

Solve.

1. $5 + (9 - 2) = x$ **2.** $(30 + 40) * 5 = z$

Add parentheses to make each statement true.

3. $30 - 12 + 5 = 13$ **4.** $60 = 4 + 6 * 6$

Check your answers in the Answer Key.

Order of Operations

To avoid confusion when solving number sentences, mathematicians have agreed to a set of rules called the **order of operations.** These rules tell what operation to do first and what to do next.

The rules are important to follow so everyone gets the same solution to a problem. For example, think about ? = 8 + 4 * 3. If one person added first and another person multiplied first, they would get different answers.

Rules for the Order of Operations

1. Do operations inside any **parentheses** first. Follow rules 2 and 3 when computing inside parentheses.
2. **Multiply** and **divide** in order, from left to right.
3. **Add** and **subtract** in order, from left to right.

Example

Solve. 8 + 4 * 3 = **?**

Multiply first.	8 + 4 * 3
Then add.	8 + 12
The answer is 20.	20

8 + 4 * 3 = **20**

Example

Solve. **?** = 10 − 6 + 2

There are no parentheses and no multiplication or division signs, so follow Rule 3.

Rule 3 says to add and subtract in order from left to right.

So subtract first.	10 − 6 + 2
Then add.	4 + 2
The answer is 6.	6

6 = 10 − 6 + 2

Example

Solve. 10 − (9 − 6 ÷ 2) = **f**

Start inside parentheses.

Divide first.	10 − (9 − 6 ÷ 2)
Then subtract.	10 − (9 − 3)
Then subtract again.	10 − 6
The answer is 4.	4

10 − (9 − 6 ÷ 2) = **4**

Did You Know?

Scientific calculators follow the rules for the order of operations. Many four-function calculators do not.

Using Variables

You can use **variables** to stand for unknown values in number sentences (equations).

Michael Phillips/E+/Getty Images

Example

Here are number sentences with different variables for unknown values.

Number sentence	_____ + 11 = 15	8 ÷ ? = 2	25 = 10 + x	y = 6 * 9
Solution	**4** + 11 = 15	8 ÷ **4** = 2	25 = 10 + **15**	**54** = 6 * 9

A **number model** is a number sentence that represents a real-world situation such as a number story. You can use a letter or other symbol to stand for the unknown value in a number model for a number story. Choosing the first letter of the unit of the unknown quantity can help you remember the unit you are trying to find.

Example

The temperature was 84 degrees at 6:00 A.M. At noon it was 100 degrees. What was the temperature change?

Use the letter t to stand for the temperature change.

A number model for this number story is

$84 + t = 100$

$84 + \mathbf{16} = 100$

The unknown value in the number story is 16.

The temperature increased by 16 degrees.

Change

Start		End
84	t	100
	_____	

Using Variables in Rules

Function machines and "What's My Rule?" tables have rules that tell you how to get the *out* numbers from the *in* numbers. For example, a doubling machine might have the rule, "Double the *in* number." This rule can be written as $y = 2 * x$ using variables.

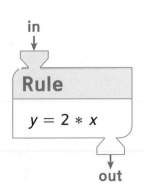

in
↓

Rule

$y = 2 * x$

↓
out

in	out
x	y
0	0
1	2
2	4
3	6

Using Variables in Formulas

A **formula** is a general rule for finding the value of something. Formulas are used in everyday life, in science, and in business. For example, the formula for the area of a rectangle is $A = l * w$, where A is the area, l is the length, and w is the width.

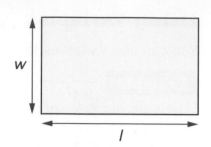

Variables Can Be Used to State Properties of the Number System

Properties of the number system are statements that are true for all numbers. For example, any number multiplied by 1 is equal to itself. Variables are often used in number sentences that describe properties.

The table below summarizes some important multiplication properties and strategies that use variables. These properties are explained in further detail on pages 39–41.

Multiplication Strategy and Related Property	Examples
Multiplying by 1 (Identity Property) $a * 1 = a$	$1 * 3 = 3$ $48 * 1 = 48$
Multiplying by 0 (Zero Property) $a * 0 = 0$	$0 * 2 = 0$ $312 * 0 = 0$
Turn-Around Rule (Commutative Property) $a * b = b * a$	$3 * 4 = 12$ $\quad$ $4 * 3 = 12$ $10 * 4 = 40$ $\quad$ $4 * 10 = 40$
Multiplying More Than Two Factors (Associative Property) $(a * b) * c = a * (b * c)$	$4 * 2 * 3 = (4 * 2) * 3 = 8 * 3 = 24$ or $4 * 2 * 3 = 4 * (2 * 3) = 4 * 6 = 24$
Break-Apart Strategy (Distributive Property) $a * b$, where $b = c + d$ $a * b = a * (c + d)$ $a * b = (a * c) + (a * d)$	$8 * 7$, breaking apart 7 into $5 + 2$ $8 * 7 = 8 * (5 + 2)$ $8 * 7 = (8 * 5) + (8 * 2)$ $8 * 7 = 40 + 16$ $8 * 7 = 56$

Note The formula $A = l * w$ can also be written without a multiplication symbol: $A = lw$. Putting variables next to each other means they are to be multiplied.

In higher mathematics, in which there are many variables, multiplication symbols are normally left out. In *Everyday Mathematics*, however, the multiplication symbol is usually shown because this makes expressions easier to understand.

Check Your Understanding

If you know a person's weight, you can use this formula to estimate the volume of blood in his or her body:

$B = W / 28$, where B is the volume of blood (in liters) and W is the person's weight (in pounds).

Estimate the volume of blood in a person who weighs 84 pounds.

Check your answer in the Answer Key.

Multiplication Properties

Certain statements are true of all numbers. Some are obvious, such as "every number equals itself." Others are less obvious. You have probably already used most of these **properties** in earlier grades, **b**ut you may not know their mathematical names.

The Identity Property of Multiplication: Multiplying by 1

The product of any number and 1 is that number: $a * 1 = a$ and $1 * a = a$.

For example, $75 * 1 = 75$ and $1 * 75 = 75$.

When you multiply a number by 1, you can think of having 1 group of that number of objects. So the product will be equal to the number you multiplied by 1.

$1 * 2$ means you have 1 group of 2 objects, which means you have 2 objects in all.

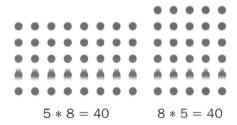

1 group of
2 shoes

The Commutative Property of Multiplication: The Turn-Around Rule for Multiplication

The **Commutative Property of Multiplication** is what you have called the **turn-around rule for multiplication** in earlier grades. It says you can multiply **factors** in any order:

$a * b = b * a$

You can see an example when you turn an array and still have the same **product.** Nothing has been added to or taken any away from the total amount.

$5 * 8 = 40$ $8 * 5 = 40$

You can think about the **Commutative Property** in everyday equal-grouping situations.

Example

Lindsay has 8 boxes of 2 pens. How many pens does she have?

$8 * 2 = 16$ She has 16 pens.

Tommy has 2 boxes of 8 pens. How many pens does he have?

$2 * 8 = 16$ He has 16 pens.

Both Lindsay and Tommy have 16 pens.

Note Using the Commutative Property to change the order of factors can make facts easier to solve. For example, if you don't know what $9 * 5$ equals, you can solve the turn-around fact by thinking about $5 * 9$. Since $5 * 9 = 45$, you know that $9 * 5 = 45$.

The Associative Property of Multiplication: Multiplying More Than Two Factors

When you multiply more than two factors together, the way you group the factors does not matter. It makes no difference which two factors you multiply first.

$(a * b) * c = a * (b * c)$

Example

$3 * 5 * 2 = ?$

- $(3 * 5) * 2 = ?$
- Multiply $3 * 5 = 15$ first.
- Then $15 * 2 = 15 + 15 = 30$.
- The product is **30**.

- $3 * (5 * 2) = ?$
- Multiply $5 * 2$ first because $5 * 2 = 10$ is a helper fact.
- Then $3 * 10 = 30$.
- The product is still **30**.

$3 * 5 * 2 = 30$

The Distributive Property

When you use **partial-products multiplication**, you are using the **Distributive Property:**

$a * (b + c) = (a * b) + (a * c)$

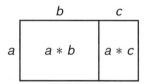

For example, when you solve $5 * 28$ using partial products, you can think of 28 as $20 + 8$ and multiply each part by 5.

The Distributive Property says $5 * (20 + 8) = (5 * 20) + (5 * 8)$.

	20	+ 8
5	5 * 20 = 100	5 * 8 = 40

```
                    2   8
    *                   5
    _____
5 * 20 →      1     0   0
5 * 8 →                 4   0
    _____
             1     4   0
```

You can use rectangles to show how the Distributive Property works.

Example

Show how the Distributive Property works by finding the area of the large rectangle in two different ways.

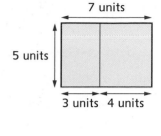

Method 1: To find the area of the large rectangle, you can find its total width (3 + 4) and multiply that by the length.

$$5 * (3 + 4) = 5 * 7$$
$$= 35 \text{ square units}$$

Method 2: Another way to find the area of the large rectangle is to first find the area of each smaller rectangle, and then add those areas.

$$(5 * 3) + (5 * 4) = 15 + 20$$
$$= 35 \text{ square units}$$

Both methods show that the area of the rectangle is 35 square units.

$$5 * (3 + 4) = (5 * 3) + (5 * 4)$$

The Distributive Property also works with subtraction:

$$a * (b - c) = (a * b) - (a * c)$$

Example

Find the area of the shaded rectangle in two different ways.

Method 1: $5 * (7 - 4) = 5 * 3 = 15$ square units

Method 2: $(5 * 7) - (5 * 4) = 35 - 20 = 15$ square units

Both methods show that the area of the shaded rectangle is 15 square units.

$$5 * (7 - 4) = (5 * 7) - (5 * 4)$$

> Describe what each number represents in each number model. For example, the 7 represents the width of the large rectangle.

Multiplication Fact Strategies

Multiplication fact strategies have helped you become fluent with multiplication facts. These strategies often use facts you know really well, called helper facts, to solve other facts. The 2s, 5s, 10s, and squares are often used as helper facts.

Adding-a-Group Strategy

Adding a group to a helper fact can help you figure out other facts. You can show the adding-a-group strategy by adding a row to an array to find the new, larger product.

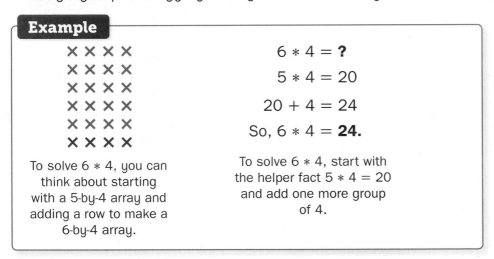

Example

$6 * 4 = ?$

$5 * 4 = 20$

$20 + 4 = 24$

So, $6 * 4 = $ **24.**

To solve $6 * 4$, you can think about starting with a 5-by-4 array and adding a row to make a 6-by-4 array.

To solve $6 * 4$, start with the helper fact $5 * 4 = 20$ and add one more group of 4.

The adding-a-group strategy works well for 3s facts and 6s facts because you can use 2s facts and 5s facts as helpers.

Subtracting-a-Group Strategy

Sometimes you can start from a helper fact and subtract a group to figure out a nearby fact. You can show the subtracting-a-group strategy by crossing out a row on an array to find the new, smaller product.

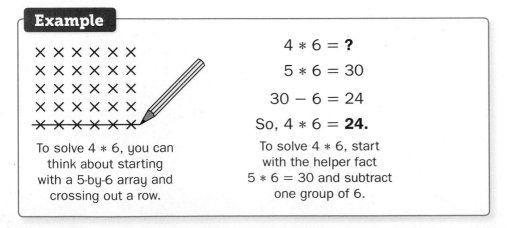

Example

$4 * 6 = ?$

$5 * 6 = 30$

$30 - 6 = 24$

So, $4 * 6 = $ **24.**

To solve $4 * 6$, you can think about starting with a 5-by-6 array and crossing out a row.

To solve $4 * 6$, start with the helper fact $5 * 6 = 30$ and subtract one group of 6.

The subtracting-a-group strategy works well for 9s facts and 4s facts because you can use 10s facts and 5s facts as helpers.

Doubling Strategy

Doubling can be helpful with facts that are doubles facts you know well. Area models can help you represent doubling. The rectangle below shows how you can solve 6 * 7 by thinking of a rectangle with side lengths of 6 feet and 7 feet. To find the area of the whole rectangle, it can be easier to think of the rectangle as two rectangles that are each 3 feet by 7 feet. Find the area of one rectangle and then double the area to solve 6 * 7.

```
              7 ft
        ┌──────────────────┐
        │ 3 ft   3 * 7     │  3 * 7 = 21  ⎫
 6 ft   ├──────────────────┤              ⎬  21 + 21 = 42
        │ 3 ft   3 * 7     │  3 * 7 = 21  ⎭  so 6 * 7 = 42 sq ft
        └──────────────────┘
```

Doubling works for facts with at least one even factor because you can start by halving that factor.

Example

4 * 7 = **?**

Think: I know half of 4 is 2, so I can start with 2 * 7 = 14. Then I can double 2 * 7 to find 4 * 7.

14 + 14 = 28, so 4 * 7 = **28.**

Near-Squares Strategy

You can solve facts that are close to a square number (such as 7 * 7 or 8 * 8) by using the near-squares strategy. You can add or subtract groups from a nearby square number.

```
x x x x x x x       x x x x x x x       x x x x x x x
x x x x x x x       x x x x x x x       x x x x x x x
x x x x x x x       x x x x x x x       x x x x x x x
x x x x x x x       x x x x x x x       x x x x x x x
x x x x x x x       x x x x x x x       x x x x x x x
x x x x x x x       x x x x x x x       x x x x x x x
x x x x x x x       x-x-x-x-x-x-x       x x x x x x x
                                        • • • • • • •
   7 * 7 = 49         6 * 7 = 42          8 * 7 = 56
```

You can start with 7 * 7 = 49 as your helper fact.	You can subtract a group from 7 * 7 to solve 6 * 7.	You can add a group to 7 * 7 to solve 8 * 7.
	(7 * 7) − 7 = 42	(7 * 7) + 7 = 56
	49 − 7 = 42	49 + 7 = 56
	6 * 7 = 42	8 * 7 = 56

The near-squares strategy works well when the factors are close together, like 6 * 7 and 8 * 7.

Diagrams for 1- and 2-Step Number Stories

You can use tools such as drawings, arrays, and number lines to help you make sense of and solve problems. Situation diagrams are another tool that can help you organize your thinking and write number models for number stories.

Examples

Rhodes Elementary School students collected books to donate to a local center. They collected 432 books in the first week and 461 books in the second week.

You can make sense of the problems below by filling in situation diagrams with information from the number story.

Problem 1: How many total books did they collect in the two weeks?

You can think of this as a parts-and-total story where two or more parts are combined to form a total. A **parts-and-total diagram** shows that the total is the same size as the two parts combined.

Let t stand for the total number of books collected in two weeks.

Total
books
t

Part	Part
books week 1	books week 2
432	461

Number model: $432 + 461 = t$
Summary number model: $432 + 461 = 893$
The students collected 893 books.

Problem 2: How many more books did they collect the second week than the first week?

You can think of this as a comparison story. A **comparison diagram** shows the difference between two quantities.

Let d stand for the difference in the number of books.

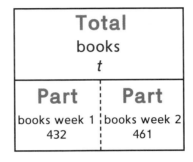

Number model: $461 - 432 = d$
Summary number model: $461 - 432 = 29$
They collected 29 more books the second week.

Situation diagrams can help you organize your thinking for 2-step number stories.

Example

There are 7 rows of children with 4 children in each row on a bus. At the first stop, 5 children get off the bus. How many children are on the bus now?

Think about what you know and what you need to find out. You can organize the amounts in a change diagram.

There are 7 rows of children with 4 children in each row. That is the starting number of children on the bus. Write (7 * 4) in the Start space.

The number of children decreases by 5 when 5 children get off the bus. Write − 5 in the Change space.

You can use the letter C to remember that you are looking for the number of <u>c</u>hildren still on the bus. Write C in the End space.

You can write a number model from the diagram: $(7 * 4) - 5 = C$.

Then solve the problem:

$(\mathbf{7 * 4}) - 5 = C$
$\mathbf{28} - 5 = C$
$23 = C$

There are 23 children on the bus. $(7 * 4) - 5 = \mathbf{23}$

You can use a multiplication/division diagram to organize 2-step problems.

Example

Ms. Fry buys 10 packages of paints for her art classes. Each package contains 4 paint trays. If she needs 5 trays for each class, how many classes will have enough paint?

She has 10 packages of 4 paint trays, or a total of (10 * 4) paint trays. She needs 5 paint trays per class.

She needs to find how many classes will have enough paint.

A number model for the story is $(10 * 4) \div 5 = \mathbf{c}$.

classes	paint trays per class	paint trays in all
c	5	(10 * 4)

$(\mathbf{10 * 4}) \div 5 = c$
$\mathbf{40} \div 5 = c$
$8 = c$

She has enough paint trays for 8 classes. $(10 * 4) \div 5 = \mathbf{8}$

Estimation

An **estimate** is an answer that should be close to an exact answer. Making an estimate when you first start working on a problem may help you understand the problem better. Estimating before you solve a problem is like making a rough draft of a writing assignment: you have an idea about the ending, but you don't have all of the details written yet.

Estimation is also useful after you have found an answer to a problem. You can use the estimate to check whether your answer makes sense. If your estimate is not close to the exact answer you found, then you need to check your work or try a different estimation strategy.

There are many ways to make a reasonable estimate. You can use front-end estimation, round numbers in the problem, or use close-but-easier numbers. You can read more about these estimation strategies and methods on pages 84–89.

Example

Mrs. Ward has 19 packages of 12 rulers. She needs 275 rulers for all the students in the school. How many more rulers does she need to order?

First Mrs. Ward makes an estimate using close-but-easier numbers. 19 is close to 20. She has about 20 * 12, or 240, rulers. She needs about 280 rulers since 275 is close to 280. 280 − 240 = 40 rulers. She estimates that she needs to order about 40 more rulers.

She organizes her problem in a comparison diagram to find out exactly how many rulers she has and how many more she needs.

She writes the number model 275 − (19 * 12) = r.

Then she solves the number sentence.

She multiplies 19 * 12 first. Then she subtracts.

Quantity
275

Quantity		
(19 * 12)		r
		_____
		Difference

Summary number model:
275 − (19 * 12) = 47
Mrs. Ward needs to order 47 rulers.
Her answer is reasonable because 47 is close to her estimate of 40 rulers.

```
        1  9
   *    1  2
   --------
        1  8
        2  0
        9  0
   + 1  0  0
   --------
     2  2  8
```

```
     2 7 5
   - 2 2 8
   -------
       4 7
```

Arrays and Factors of Counting Numbers

When two numbers are multiplied, the answer is called the **product.** The two numbers that are multiplied are called **factors** of the product.

Note Factors of counting numbers *must* be counting numbers. **Counting numbers** are the numbers 1, 2, 3, 4, 5, and so on.

Examples

Name the factors and the products.

$$3 * 6 = 18$$
factors   product

$$1 * 18 = 18$$
factors   product

$$480 = 40 * 12$$
product   factors

An **array** is a group of objects arranged in **rows** and **columns.** An outline around the rows and columns would be a rectangle. Each row is filled and has the same number of objects. Each column is filled and has the same number of objects.

Arrays can show all the factors of a counting number.

The keys on the phone are in a 4-by-3 array.

Example

Find all of the factors of 6. Use a system to organize your thinking.

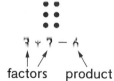

$$1 * 6 = 6$$
factors   product

6 dots can form an array of 1 row with 6 dots in the row. 1 and 6 are factors of 6.

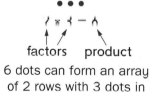

$$2 * 3 = 6$$
factors   product

6 dots can form an array of 2 rows with 3 dots in each row. 2 and 3 are factors of 6.

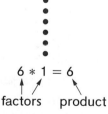

$$3 * 2 = 6$$
factors   product

Because of the turn-around rule, 6 dots can form an array of 3 rows with 2 dots in each row. We already found that 3 and 2 are factors of 6.

6 dots cannot form an array of 4 equal rows. 4 is *not* a factor of 6. 4 cannot be paired with another counting number to make a product of 6.

6 dots cannot form an array of 5 equal rows. 5 is *not* a factor of 6. 5 cannot be paired with another counting number to make a product of 6.

$$6 * 1 = 6$$
factors   product

We already showed that 6 and 1 are factors of 6 in the 1-by-6 array. Because of the turn-around rule, we did not need to rearrange the dots here.

The arrays show that the factors of 6 are 1, 2, 3, and 6. These are the *only* factors of 6.

Kinds of Counting Numbers

The **counting numbers** are the numbers 1, 2, 3, 4, 5, and so on.

A counting number is an **even number** if 2 is one of its factors. That means an even number can be divided by 2 with no remainder. The even numbers are 2, 4, 6, 8, 10, 12, and so on.

A counting number is an **odd number** if 2 is *not* one of its factors. An odd number divided by 2 will always have a remainder of 1. The odd numbers are 1, 3, 5, 7, 9, 11, and so on.

A **prime number** is a counting number that has exactly 2 different factors. For example, 5 is a prime number because the only factors of 5 are 1 and 5. All of these numbers are prime numbers:

2, 3, 5, 7, 11, 13, 17, 19

A **composite number** is any counting number that has 3 or more different factors. These are all counting numbers greater than 1 that are not prime numbers. All of these numbers are composite numbers:

4, 6, 8, 9, 10, 12, 14, 15, 16, 18, 20

Facts about the Numbers 1 through 20			
Number	Factors	Prime or Composite	Even or Odd
1	1	neither	odd
2	1 and 2	prime	even
3	1 and 3	prime	odd
4	1, 2, and 4	composite	even
5	1 and 5	prime	odd
6	1, 2, 3, and 6	composite	even
7	1 and 7	prime	odd
8	1, 2, 4, and 8	composite	even
9	1, 3, and 9	composite	odd
10	1, 2, 5, and 10	composite	even
11	1 and 11	prime	odd
12	1, 2, 3, 4, 6, and 12	composite	even
13	1 and 13	prime	odd
14	1, 2, 7, and 14	composite	even
15	1, 3, 5, and 15	composite	odd
16	1, 2, 4, 8, and 16	composite	even
17	1 and 17	prime	odd
18	1, 2, 3, 6, 9, and 18	composite	even
19	1 and 19	prime	odd
20	1, 2, 4, 5, 10, and 20	composite	even

Note There are only 2 ways to make an array with 5 dots. I and 5 are the only factors of 5, so 5 is a prime number.

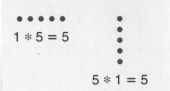

$1 * 5 = 5$

$5 * 1 = 5$

Note Prime and composite numbers have at least two different factors. The number I has only one factor. So the number I is neither prime nor composite.

Did You Know?

Many people are interested in identifying large prime numbers. See http://primes.utm.edu/largest.html for details and for lists of large and small prime numbers.

Multiples

A **multiple** of a number is the product of that number and another counting number. For example, 15 is a multiple of 5 because it is a product of 5 and 3. For the multiples of 5 (5, 10, 15, 20, and so on) think about multiplication:

$1 * 5 = 5$ $2 * 5 = 10$ $3 * 5 = 15$ $4 * 5 = 20$

Note *Multiple* is another name for the product of 2 numbers:

factor * factor = product

factor * factor = multiple

A number is a multiple of each of its factors.

Examples

Find multiples of 2, 3, and 5.

Multiples of 2: 2, 4, 6, 8, 10, 12, 14, 16, 18, 20, 22, 24, …

Multiples of 3: 3, 6, 9, 12, 15, 18, 21, 24, 27, 30, 33, 36, …

Multiples of 5: 5, 10, 15, 20, 25, 30, 35, 40, 45, 50, 55, …

Note When you skip-count by a number, your counts are the multiples of that number. Since you can always count further, lists of multiples go on forever.

Common Multiples

When listing multiples for two or more numbers, you may find some of the same numbers are on all the lists. These numbers are called **common multiples.**

Example

Find common multiples of 2 and 3.

Multiples of 2: 2, 4, **6**, 8, 10, **12**, 14, 16, **18**, 20, 22, **24**,

Multiples of 3: 3, **6**, 9, **12**, 15, **18**, 21, **24**, 27, …

Common multiples of 2 and 3: **6, 12, 18, 24, …**

Example

How are the factors of 10 related to the multiples of the factors of 10?

Make a table of the factors of 10 and multiples of the factors of 10 up to 10.

Find the common multiple of all the factors.

Factors of 10	Multiples of the Factors of 10
1	1, 2, 3, 4, 5, 6, 7, 8, 9, **10**
2	2, 4, 6, 8, **10**
5	5, **10**
10	**10**

10 is a multiple of every factor of 10. So, 10 is a common multiple of all its factors.

Check Your Understanding

Find two common multiples for each pair of numbers.

1. 6 and 12   **2.** 4 and 10   **3.** 9 and 15

Check your answers in the Answer Key.

Multiplicative Comparisons

A comparison statement is a statement that compares two or more items or amounts. For example, "Mia is taller than Jasmine." In mathematics, we often compare different amounts, or *quantities*, using terms such as *greater than* or *less than*.

You can make an **additive comparison statement** that describes how numbers are related in terms of addition. Or, you can make a **multiplicative comparison statement** that considers how quantities relate in terms of multiplication.

Examples

An additive comparison:

> Robert bought 6 comic books.
> Kat bought 7 more than Robert.
> Kat's comic books = 6 + 7

So Kat bought 13 comic books.

A multiplicative comparison:

> Robert bought 6 comic books.
> Kat bought 7 times as many as Robert.
> Kat's comic books = **7** * 6

So Kat bought 42 comic books.

When thinking about multiplicative comparison statements, ask "How many times as much?" or "How many times as many?"

Example

Make a multiplicative comparison statement comparing the length of a 6-inch string and a 2-inch string.

6 inches

2 inches

You know the 6-inch string is longer than the 2-inch string. To make a multiplicative comparison statement, ask: *How many times longer is the 6-inch string compared to the 2-inch string?*

Multiplicative comparison statements for the two lengths of string include:

- The 6-inch string is 3 times as long as the 2-inch string.
- The 6-inch string is 3 times the length of the 2-inch string.
- The 6-inch string is 3 times longer than the 2-inch string.

Note Equations and number models can show the relationship between two quantities. In this example, an equation showing the relationship between the lengths of the strings is $6 = 3 * 2$. The 6 represents the length of the longer string and the 2 represents the length of the shorter string. The 3 shows the relationship between the two lengths: the longer string is 3 times the length of the shorter string.

When writing equations showing multiplicative comparisons, each equation involves two quantities (one of the factors and the product) and another number (the other factor) that represents *how many times as much* or *how many times as many*. You can write two comparison statements for this equation: $18 = 3 * 6$.

- 18 is 3 times as much as 6.
- 18 is 6 times as much as 3.

Number models represent real-world situations. You can model comparison stories with equations.

Example

Write a comparison story that matches the equation $18 = 3 * 6$ and the comparison statement *18 is 3 times as much as 6*.

Ainsley has 6 action figures. Her older brother Jordan has 3 times as many action figures. How many action figures does Jordan have?
Jordan has 18 action figures.

Sometimes it is helpful to draw a diagram to represent the problem.

Example

Patrick hiked 20 minutes. Mabel hiked 3 times as long. How long did Mabel hike?

An equation representing this problem is $20 * 3 = h$.
The solution to this equation is 60. Mabel hiked 60 minutes

Patrick | 20 min |

Mabel | 20 min | 20 min | 20 min |

Example

Derek rode his bicycle 9 miles on Saturday. Rachel rode her bicycle 36 miles on Saturday. How many times farther did Rachel ride than Derek?

You have to find out how many 9s are in 36. $n * 9 = 36$
There are four 9s in 36. $4 * 9 = 36$
Rachel rode 4 times as far as Derek.

D | 9 |
R | 36 |

Check Your Understanding

Write an equation that matches each multiplicative comparison statement.

1. 32 is 4 times as much as 8.   **2.** What number is 7 times as much as 5?

Write an equation that matches the number story. Then solve.

3. Ms. Rettig planted 3 times as many tomato plants this year as she did last year. She planted 5 tomato plants last year. How many did she plant this year?

Check your answers in the Answer Key.

Example

The rule and the *in* numbers are given. Find the *out* numbers.

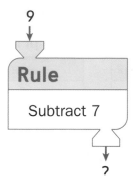

9

Rule

Subtract 7

?

in	out	
9	2	[9 − 7 = 2]
29	22	[29 − 7 = 22]
59	52	[59 − 7 = 52]
79	72	[79 − 7 = 72]

The solutions (the missing information) appear in color.
What do you notice about the digits in the ones place? When 7 is subtracted from a number with a 9 in the ones place, the answer has a 2 in the ones place.

What patterns do you notice in the "What's My Rule?" tables that aren't stated in the rule?

If you know the rule and the *out* numbers, then you should be able to find the *in* numbers.

Example

The rule is " − 5." Find the numbers that were put into the machine.

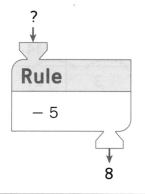

?

Rule

− 5

8

in	out
	8
	48
	88

This machine subtracts 5 from any number. The number that comes out is always 5 less than the number put in.
If 8 comes out, then 13 was the number put in. 13 − 5 = 8
If 48 comes out, then 53 was the number put in. 53 − 5 = 48
If 88 comes out, then 93 was the number put in. 93 − 5 = 88

Example

The rule and the *out* numbers are given. Find the *in* numbers.

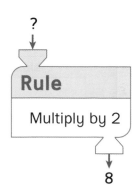

?

Rule

Multiply by 2

8

in	out	
4	8	[4 * 2 = 8]
24	48	[24 * 2 = 48]
$\frac{1}{2}$	1	[$\frac{1}{2}$ * 2 = 1]
0	0	[0 * 2 = 0]

Example

The rule and some *in* and *out* numbers are given. Find the missing numbers.

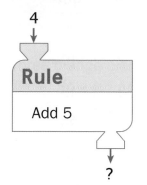

in	out
4	9
7	12
53	58
99	104

[4 + 5 = 9]

[7 + 5 = 12]

[53 + 5 = 58]

[99 + 5 = 104]

The solutions (the missing information) appear in color.

What patterns do you see in the table? Look across the rows and down the columns.

Example

Use the table to find a rule.

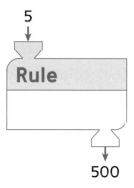

in	out
6	600
8	800
10	1,000
12	1,200

A rule for this table is "* 100."

Example

The *in* and *out* numbers are given. Find a rule.

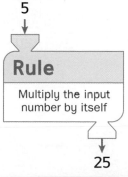

in	out
2	4
5	25
1	1
10	100

[2 * 2 = 4]

[5 * 5 = 25]

[1 * 1 = 1]

[10 * 10 = 100]

The solution (the missing information) appears in color.

Check Your Understanding

Find the missing rules. Then find the missing *in* and *out* numbers.

1.

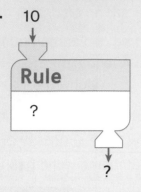

in	out
1	4
2	5
3	6
10	?
?	18

2.

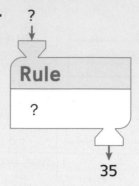

in	out
1	5
2	10
3	15
?	35
9	?

Solve these "What's My Rule?" problems.

3.

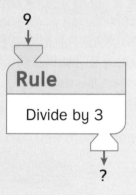

in	out
9	?
36	?
1	?
210	?
123	?
390	?

4.

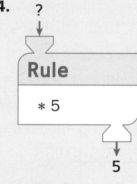

in	out
?	5
?	15
?	20
?	35
?	40
?	55

5.

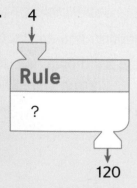

in	out
4	120
10	300
0	0
33	990
14	420
?	600

6. What other patterns do you notice in the "What's My Rule?" table in Problem 4 that are not stated in the rule "＊5"?

Check your answers in the Answer Key.

Sound, Music, and Mathematics

Musicians make patterns of sound to create music. Mathematics can help you understand how both sound and music are created.

Sound

Every sound you hear begins with a vibration—a back and forth motion. For musical instruments to produce sound, something must be set in motion.

Faster vibrations produce higher-pitched notes, and slower vibrations produce lower-pitched notes. The speed at which a string, a drumhead, or a column of air vibrates is called the *frequency*. Frequency is measured in Hertz (Hz), or "vibrations per second." The human ear can hear vibrations from about 15 Hz to 20,000 Hz.

The sound of a drum starts when a person beats the drumhead. When the drumhead stops vibrating, the sound stops.

The sound of a guitar starts when a person plucks or strums the strings. Each vibrating string moves back and forth at the same rate until it stops moving. When the strings stop vibrating, the sound stops.

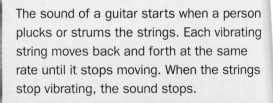

The sound of a flute starts when a person blows across the mouthpiece. A column of air moves back and forth inside the flute. When the player stops blowing, the column of air stops vibrating and the sound of the flute stops.

Instrument Length and Pitch

Many instruments rely on a vibrating column of air to make sound. A longer column of air vibrates at a lower frequency and makes a deeper, or lower-pitched, note. Shorter vibrating air columns make higher-pitched notes.

Here are some instruments you may have heard, along with the frequency of the lowest note that can be played on the instrument. What happens to the frequency as the instruments get shorter?

bassoon, 58 Hz

clarinet, 139 Hz

oboe, 233 Hz

piccolo, 587 Hz

A recorder can play a range of pitches. By covering all of the finger holes on a recorder, the musician creates the longest possible column of air and the lowest-pitched note. With all holes uncovered, a high note is produced.

The piccolo has a very short column of air within it, so it produces high-pitched notes. Piccolos produce notes in the range of about 600 to 4,000 Hz.

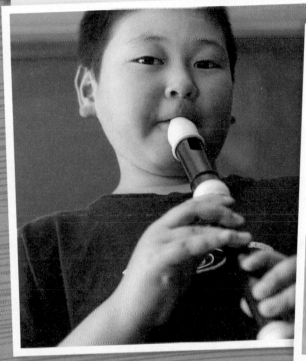

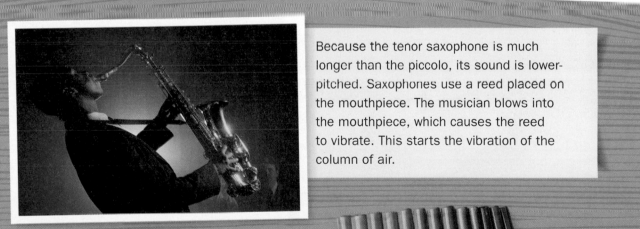

Because the tenor saxophone is much longer than the piccolo, its sound is lower-pitched. Saxophones use a reed placed on the mouthpiece. The musician blows into the mouthpiece, which causes the reed to vibrate. This starts the vibration of the column of air.

This Andean panpipe is played by blowing across the edges of hollow tubes of different lengths. The player slides the instrument from side to side to change notes. Which tube do you think would make the highest pitch? Which would make the lowest pitch?

Percussion Instruments

Percussion instruments, such as drums, produce sound by being struck, scraped, or shaken. The size of an instrument affects the pitch it can play. The size and tightness of the drumhead and the materials that the drumhead is made from also affect the pitch.

A drummer holds a West African talking drum, or *donno*, between the upper arm and the body. Squeezing the strings with the upper arm tightens the drumhead and raises the pitch of the drum. Releasing the strings loosens the drumhead and lowers the pitch of the drum.

In a trap set, the largest drum—the bass or "kick" drum—produces the lowest-pitched notes. Each drum can be tuned up or down by tightening or loosening the heads.

The steel drum, also known as a steelpan, is from the Caribbean island of Trinidad and is made by cutting off the top of a steel oil barrel. Each small rounded section of the drumhead is shaped to play a different pitched note. The pitch of the instrument can be very high because the small metal sections vibrate rapidly.

Stringed Instruments

The pitch of the notes that a stringed instrument can play is related to the length, thickness, and tension of the strings.

The violin has short, thin strings. It is designed to play higher-pitched notes. When a player presses down on a string on the fingerboard, the vibrating part is shortened and the pitch becomes higher.

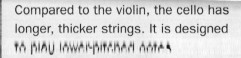

Compared to the violin, the cello has longer, thicker strings. It is designed to play lower-pitched notes.

This man is tuning his guitar by changing the tension of the strings. Tightening a string makes the pitch higher. Loosening a string makes the pitch lower.

The Piano

Looking closely at the way a piano works can help you see some of the mathematical relationships in music.

A piano's sound begins when a player presses a key. This causes a felt-covered wooden hammer to hit the strings for that key. The strings then vibrate to produce sound. Each key produces a note with a different pitch.

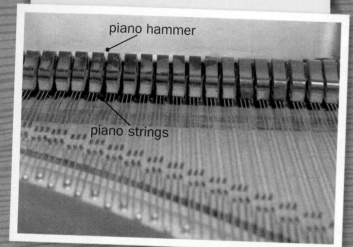

piano hammer

piano strings

An octave begins and ends on a note with the same name. For example, the keys between "Middle C" and the C to the right of it represent one octave. There are eight octaves on most pianos.

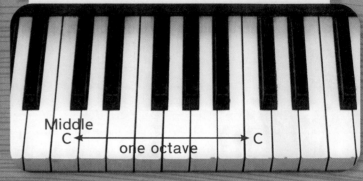

Middle C ←— one octave —→ C

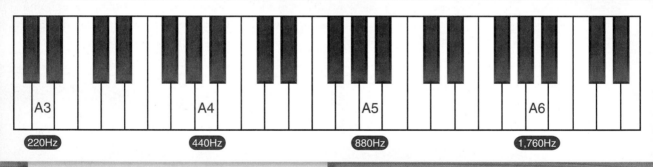

A3   A4   A5   A6

220Hz   440Hz   880Hz   1,760Hz

As you move to the right on the piano keyboard, the frequencies get higher. What patterns do you see in the frequencies?

What patterns can you find in music? How have you seen mathematics used in music?

This tuning fork vibrates 440 times per second. Piano tuners used to tighten or loosen the A4 string until its pitch exactly matched the pitch of the vibrating tuning fork. Then all other strings were tightened or loosened based on that note. Now piano tuners tighten or loosen strings to match the sound of electronic tuners.

Number and Operations in Base Ten

Uses of Numbers

It is hard to live even one day without using or thinking about numbers. Numbers are used on clocks, calendars, car license plates, rulers, scales, and so on. The major ways that numbers are used are listed below.

Malibu
CITY LIMIT
POP. 15,272   ELEV. 20

• Numbers are used for counting.

Examples

The first U.S. Census, taken in 1790, counted 3,929,214 people.
The population of Malibu is 15,272.

• Numbers are used for measuring.

Examples

Ava swam the length of the pool in 40.23 seconds.
The package is 25 inches long and weighs $3\frac{1}{4}$ pounds.

• Numbers are used to show where something is in a reference system.

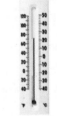

Examples

Situation	Type of Reference System
Normal room temperature is 22°C.	Celsius temperature scale
Addison was born on May 25, 2006.	Calendar
The time is 6:00 P.M.	Clock time
Detroit is located at 42°N and 83°W.	Earth's latitude and longitude system

• Numbers are used to compare counts or measures.

Examples

There were 3 times as many boys as girls at the game.
The cat weighs $\frac{1}{2}$ as much as the dog.

• Numbers can be used for identification and as codes.

Examples

driver's license number: M286-423-2061
ZIP code: 60637   phone number: (709) 555-1212

7 80021 308088

(t)Fotosearch/Getty Images; (tc)©Pixtal/SuperStock; (bc)Thomas Northcut/Photodisc/Getty Images; (b)Michael Burrell/Alamy

Kinds of Numbers

The **counting numbers** are the numbers used to count things. The set of counting numbers is 1, 2, 3, 4, and so on.

The **whole numbers** are any of the numbers 0, 1, 2, 3, 4, and so on. The whole numbers include all of the counting numbers and zero (0).

Counting numbers are useful for counting, but they do not always work for measures. Some measures fall between two whole numbers. **Fractions** and **decimals** were invented to keep track of such measures.

Fractions are often used in recipes for cooking and for measurements in carpentry and other building trades. Decimals are used for almost all measures in science and industry. Money amounts are usually written as decimals.

Examples

The recipe calls for $2\frac{3}{4}$ cups of flour.

The windowsill is 2 feet $5\frac{1}{2}$ inches above the floor.

The banana weighs 5.8 ounces.

Cups for measuring 1 cup, $\frac{1}{2}$ cup, $\frac{1}{3}$ cup, and $\frac{1}{4}$ cup

Negative numbers are used to describe numbers that are less than zero.

Examples

A temperature of 10 degrees below zero is written as −10°F.

A depth of 235 feet below sea level is written as −235 feet.

Negative numbers are also used to indicate changes in quantities.

A food scale showing weight as a decimal.

Examples

A weight loss of $4\frac{1}{2}$ pounds is recorded as $-4\frac{1}{2}$ pounds.

A decrease in income of $1,000 is recorded as −$1,000.

Geoffrey Holman/iStock/Getty Images Plus/Getty Images

Place Value for Whole Numbers

Any number, no matter how large or small, can be written using one or more of the **digits** 0, 1, 2, 3, 4, 5, 6, 7, 8, and 9. The **place value** of a number tells how much a digit is worth according to its position in a number. A place-value chart is an organizational tool used to show how much each digit in a number is worth. The **place** for a digit is its position in the number. The **value** of a digit is how much it is worth according to its place in the number.

Right to Left in the Place-Value Chart

Study the place-value chart below. Look at the numbers that name the places. As you move from right to left across the chart, each number is **10 times as large as the number to its right.**

| | * 10 | | | * 10 | | | * 10 | | | * 10 | |

10,000s ten thousands	1,000s thousands	,	100s hundreds	10s tens	1s ones

1 [10]	=	10 [1s]	10	=	10 * 1
1 [100]	=	10 [10s]	100	=	10 * 10
1 [1,000]	=	10 [100s]	1,000	=	10 * 100
1 [10,000]	=	10 [1,000s]	10,000	=	10 * 1,000

Left to Right in the Place-Value Chart

As you move from left to right across the place-value chart, each number is $\frac{1}{10}$ **the size of the number to its left.** Finding $\frac{1}{10}$ of a number is the same as dividing that number by 10.

| | ÷ 10 | | | ÷ 10 | | | ÷ 10 | | | ÷ 10 | |

10,000s ten thousands	1,000s thousands	,	100s hundreds	10s tens	1s ones

1,000	=	10,000 ÷ 10
100	=	1,000 ÷ 10
10	=	100 ÷ 10
1	=	10 ÷ 10

Example

10,000s ten thousands	1,000s thousands	,	100s hundreds	10s tens	1s ones
8	1	,	9	0	3

The number 81,903 is shown in the place-value chart above.

The value of the 8 is 80,000   or   8 [10,000s].
The value of the 1 is 1,000   or   1 [1,000].
The value of the 9 is 900   or   9 [100s].
The value of the 0 is 0   or   0 [10s].
The value of the 3 is 3   or   3 [1s].

81,903 is read as "eighty-one thousand, nine hundred three."

In larger numbers, look for commas that separate groups of 3 digits. The commas help identify the thousands, millions, and so on.

Example

The 2010 U.S. Census counted 308,745,538 people.

millions			,	thousands			,	ones		
100	10	1	,	100	10	1	,	100	10	1
3	0	8	,	7	4	5	,	5	3	8

Read from left to right. Read "million" at the first comma and "thousand" at the second comma.

The number is read as "308 **million,** 745 **thousand,** 538."

Check Your Understanding

Read each number to yourself. What is the value of the 5 in each number?

1. 35,104   **2.** 71,504   **3.** 3,657,000   **4.** 82,500,000

Check your answers in the Answer Key.

Expanded Form

The number 27,364 is written in **standard form,** the most common way of writing a number. Sometimes this is called **standard notation.**

The number 27,364 is shown in the place-value chart below. It is read "twenty-seven thousand, three hundred sixty-four."

10,000s ten thousands	1,000s thousands	,	100s hundreds	10s tens	1s ones
2	7	,	3	6	4

When a number is written as the sum of the values of each digit, it is written in **expanded form.**

The value of the 2 is 20,000. 20, 0 0 0

The value of the 7 is 7,000.   7, 0 0 0

The value of the 3 is 300.   3 0 0

The value of the 6 is 60.   6 0

The value of the 4 is 4. +   4
                    _____
              2 7, 3 6 4

In expanded form, 27,364 = 20,000 + 7,000 + 300 + 60 + 4.

There are different ways to represent a number in expanded form.

Examples

Write 54,032 in expanded form in two ways.

Write the number as the sum of the values of each digit.
54,032 = 50,000 + 4,000 + 0 + 30 + 2

Write the number using words to show the place names.
5 ten thousands + 4 thousands + 0 hundreds + 3 tens + 2 ones

Check Your Understanding

Write each number in standard form.

1. 2 [100,000s] + 3 [1,000s] + 7 [100s] + 6 [10s] + 2 [1s]

2. 5,000,000 + 200,000 + 80,000 + 9,000 + 300 + 40 + 1

Write each of the following numbers in expanded form.

3. 8,744

4. 1,456,900

Check your answers in the Answer Key.

Comparing Numbers and Amounts

When two numbers or amounts are compared, there are two possible results: They are equal because they have the same value, or they are not equal because one is larger than the other. Different symbols show that numbers and amounts are equal or not equal.

- Use an *equal sign* (=) to show that the numbers or amounts are *equal*.

- Use a *greater-than symbol* (>) or a *less-than symbol* (<) to show that they are *not equal* and to show which is larger.

Note To show that two amounts are not equal without showing which is larger, the *not-equal symbol* (≠) can be used.

Example

Symbol	=	>	<
Meaning	"equals" or "is the same as"	"is greater than"	"is less than"
Examples	$20 = 4 * 5$	$7 > 3$	$2 < 4$
	3 cm = 30 mm	352,998 > 349,999	398 < 102,045
	$5 + 5 = 6 + 6 - 2$	14 ft 7 in. > 13 ft 11 in.	99 minutes < 2 hours
	$\frac{1}{2} = \frac{5}{10}$	$8 + 7 > 9 + 5$	$2 * 6 < 4 * 5$

Before you compare amounts with units, you can write both amounts using the same unit.

Examples

Compare 30 yards and 60 feet.

The units are different—yards and feet.

1 yard

1 foot

Change yards to feet, and then compare. 1 yd = 3 ft, so 30 yd = 30 * 3 ft, or 90 ft.

Now compare feet. 90 ft > 60 ft

So, 30 yd > 60 ft.

Check Your Understanding

True or false?

1. $9 + 7 < 12$ **2.** $5 * 4 = 80 ÷ 4$ **3.** $13 + 1 > 15 - 1$

Check your answers in the Answer Key.

Estimation

An **estimate** is a number that is close to the exact answer. You make estimates every day.

- You estimate how long it will take you to walk or ride from one place to another.

- You estimate how much money you can save by the end of the year.

- You estimate how long it will take you to complete a school assignment or chore at home.

Useful estimates are based on reasoning. Estimates can help you make sense of a situation or help you approximate a calculation.

Examples

The school play will be performed in 1 month, 3 weeks, and 4 days.

When a girl says, "The play will be performed in 2 months," she's using an estimate based on knowing that there are about 4 weeks in a month.

A boy receives an allowance of $10 each week. He wonders how much money he will receive in one year.

The boy knows there are 52 weeks in a year. He mentally calculates $10 * 50 is $500. Using estimation, the boy figures out that he will receive about $500 in one year.

Sometimes estimates are used to make sense of a situation when the exact number cannot be known. For example, officials estimate the size of a crowd at a large outdoor event such as a festival or street fair. It is impossible to know exactly how many people attended the event because people are moving about and come and go at different times.

"Organizers expect over 50,000 people at the street fair."

Check Your Understanding

1. Estimate how long it takes you to get to school each day. Is it the same every day?

2. Estimate how long it will take you to do your homework tonight.

3. Estimate the number of people that can eat in your school lunchroom.

Estimation in Problem Solving

Estimation is also useful when you need to find an exact answer. You can make an estimate when you start working on a problem to help you understand the problem better. You can also use your estimate after you calculate to check whether your answer makes sense. If your answer is not close to your estimate, then you need to use another method of estimation or check your work, correct it, and check whether the new answer makes sense.

Example

Data for advanced ticket sales for the high school play showed that 249 tickets were sold for Thursday, 345 tickets were sold for Friday, and 185 tickets were sold for Saturday. How many tickets were sold in advance for the play?

249 + 345 + 185 = ?

Estimate: 249 is close to 250, 345 is close to 350, and 185 is close to 200.

   250 + 350 + 200 = 800

   About 800 tickets were sold in advance.

Calculate:

$$\begin{array}{r} \overset{1}{2}\,4\,9 \\ 3\,4\,5 \\ +\ \ 1\,8\,5 \\ \hline \cancel{6\,7\,9} \end{array}$$

Check: The numbers used to make the estimate were very close to the original numbers.

So, 679 seems too far away from the estimate of 800. Check the work again.

Correct the work:

$$\begin{array}{r} \overset{1}{2}\,\overset{1}{4}\,9 \\ 3\,4\,5 \\ +\ \ 1\,8\,5 \\ \hline 7\,7\,9 \end{array}$$

Check again: There was an error in the first calculation.

779 tickets were sold in advance.

779 seems more reasonable because it is closer to the estimate of 800 tickets.

Front-End Estimation

One way to make an **estimate** is to keep the digit in the highest place value and replace the rest of the digits with zeros. This is called **front-end estimation.**

Note Sometimes *front-end estimation* is called *leading-digit estimation.*

Examples

At its farthest, Earth is 405,500 kilometers away from the moon.

A boy estimates that Earth is about 400,000 kilometers from the moon.

The digit in the highest place value in 405,500 km is the 4 in the hundred-thousands place.

The front-end estimate is 400,000 km.

You can also estimate answers to calculations by using front-end estimation.

Example

Estimate the area of the rectangle. Use front-end estimation.

The width is about 10 cm, and the length is about 60 cm.

62 cm

13 cm

So the estimate for the area of the rectangle is about 10 * 60 or 600 sq cm.

Whenever you use front-end estimation in addition or multiplication of whole numbers, your estimate is always less than the exact answer. This is because all of the values of the numbers in the problem decrease when the digits other than the front end are replaced with zeros.

Example

Erica added 226 + 538 and got 664. Was she correct?

$$226 + 538 = ?$$
$$\downarrow \qquad \downarrow$$

Find front-end estimates for each number in the problem and add:

$$200 + 500 = 700$$

The estimate is 700. Since Erica used front-end estimation with addition, the estimate will be less than the exact answer. Erica's answer of 664 must be incorrect. She should check her work.

Note Many good estimators use front-end estimation and then get a closer estimate by using the other digits. In the example to the left, the remaining digits 26 and 38 might help Erica see that the answer should be at least 20 + 30, or 50, more than 700, or 750.

Front-end estimates can be useful for checking the reasonableness of exact calculations. If the estimate and the exact answer are not close, you should look for a mistake in your work. You may have made a mistake in your computation, or you might be able to find a closer estimate.

Rounding

Rounding is another way to make sense of numbers in situations and to estimate an answer to a calculation. Whole numbers are often rounded to the nearest multiple of 10, 100, 1,000, and so on. You can round by using or thinking about number lines.

Rounding Using Number Lines

Example

Round 4,186 to the nearest hundred.

Step 1 Draw a number line with 3 tick marks and a blank below each tick mark.

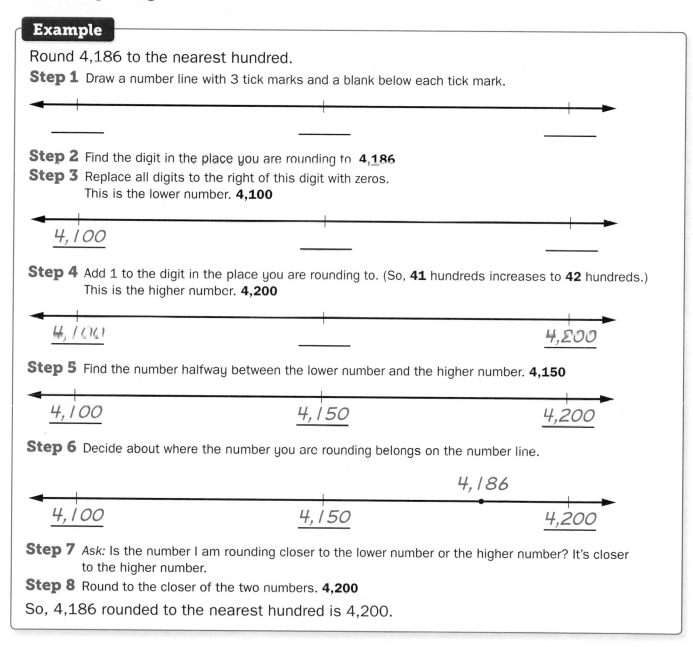

Step 2 Find the digit in the place you are rounding to 4,1_86
Step 3 Replace all digits to the right of this digit with zeros.
This is the lower number. **4,100**

Step 4 Add 1 to the digit in the place you are rounding to. (So, **41** hundreds increases to **42** hundreds.)
This is the higher number. **4,200**

Step 5 Find the number halfway between the lower number and the higher number. **4,150**

Step 6 Decide about where the number you are rounding belongs on the number line.

Step 7 *Ask:* Is the number I am rounding closer to the lower number or the higher number? It's closer to the higher number.
Step 8 Round to the closer of the two numbers. **4,200**

So, 4,186 rounded to the nearest hundred is 4,200.

The next two examples show a second method for rounding a number. It is similar to the way you think when you use a number line to round.

Example

Round 7,358 to the nearest thousand.

1. Write the number you are rounding and mark the digit in the place you are rounding to.	7,358
2. Replace all numbers to the right of this digit with 0. This is the lower number.	7,000
3. Increase the digit in the place you are rounding to by 1. This is the higher number. Note that in this case you are increasing the number by 1 thousand.	8,000
4. *Ask:* Is the number I am rounding closer to the lower number or the higher number?	lower
5. Round to the closer of those two numbers.	7,000

7,358 is closer to 7,000 than it is to 8,000.

7,358 rounded to the nearest thousand is 7,000.

Note When finding the higher number, look at the digit you are rounding to and the digits to the left of it. Say this number and increase it by one. For example, if rounding 3,924 to the nearest hundred, the lower number becomes **39** hundreds (**3,9**00). The higher number increases by one hundred to **40** hundreds (**4,0**00).

Example

Round 7,358 to the nearest hundred.

1. Write the number you are rounding and mark the digit in the place you are rounding to.	7,358
2. Replace all numbers to the right of this digit with 0. This is the lower number.	7,300
3. Increase the digit in the place you are rounding to by 1. This is the higher number. Note that in this case you are increasing the number by 1 hundred.	7,400
4. *Ask:* Is the number I am rounding closer to the lower number or the higher number?	higher
5. Round to the closer of those two numbers.	7,400

7,358 is closer to 7,400 than it is to 7,300.

7,358 rounded to the nearest hundred is 7,400.

A Rounding Shortcut

Once you practice rounding numbers by using or thinking about number lines or tables, you may notice patterns about when to round down and when to round up. The example below shows a third method for rounding numbers that is a shortcut.

Example

Round 2,478 to the nearest hundred.

- Find the digit in the hundreds place.   2,478
- Look at the digit to the right of the hundreds place.
- If the digit is less than 5, round down.
  If the digit is 5 or greater, round up.

Think: Is 7 equal to or greater than 5?

7 > 5, so round up to 2,500.   2,500

So, 2,478 rounded to the nearest hundred is 2,500.

If the number is exactly halfway between the lower and higher numbers, use the real-world information or problem situation to decide whether to round up or down. When there is no real-world information or there is nothing in the problem situation to help you decide, round to the higher number.

Check Your Understanding

First round each number to the nearest thousand. Then round each number to the nearest hundred.

1. 123,456   **2.** 12,726   **3.** 2,910

Check your answers in the Answer Key.

Close-But-Easier Numbers

When calculating, some numbers are easier to compute mentally than others. When you use **close-but-easier numbers,** you look for numbers that are close to the original number, but are easier to work with, such as numbers that end in 0 or 5 or are multiples 10 or 25. Sometimes these types of numbers are called **friendly numbers.** Selecting close-but-easier numbers can help you estimate the answers for calculations.

Example

Estimate: 543 + 328 + 112 = ?

Write the addends as close-but-easier numbers:

543 is close to 550. (50 is easy to add because it ends in 0.)

328 is close to 325. (25s are easier to add because I can think of quarters that are worth 25¢.)

112 is close to 100. (100 is easy to add.)

Mentally calculate: 550 + 325 + 100 = 975

543 + 328 + 112 is about 975.

There are different ways to make mental calculations. One way to add in your head is to group numbers that are easy for you to add together.

Example

Mentally calculate: 550 + 325 + 100 = ?

First you can add the hundreds: 500 + 300 + 100 = 900

Then you can add 50 and 25: 50 + 25 = 75

900 + 75 = 975

Can you think of other ways to mentally calculate the sum?

Example

Jada is saving to buy a tablet computer that costs $349. She saved $102 so far. Her brother gave her another $46. About how much more money does Jada need?

Write the numbers as close-but-easier numbers:

349 is close to 350.
102 is close to 100.
46 is close to 50.

Mentally calculate: 350 − 100 − 50 = 200

Jada needs about $200 more to buy the tablet computer.

Example

Jacob's family car can go 35 miles on one gallon of gas. If the car's gas tank holds 18 gallons, about how many miles can they travel on a tank of gas?

Write the numbers as close-but-easier numbers:

35 is close to 30.

18 is close to 20.

Mentally calculate: 30 * 20 = 600

The family can travel about 600 miles.

It is often helpful to think about all of the numbers in a problem before choosing the close-but-easier numbers.

Example

Estimate the quotient: 152 / 7 = ?

One way:

Write the numbers as close-but-easier numbers:

152 is close to 140. (140 is easily divided by 7.)

7 (Keep 7 the same.)

Mentally calculate: 140 / 7 = 20

Another way:

Write the numbers as close-but-easier numbers:

152 is close to 160.

7 is close to 8. (160 is more easily divided by 8 than 7.)

Mentally calculate: 160 / 8 = 20

152 / 7 is about 20.

Note In the example to the left, thinking about the problem situation can help you choose close-but-easier numbers. For this problem, it makes sense to choose two numbers to multiply for the estimate so that one number is less than one of the factors (30 is less than 35) and one number is greater than the other factor (20 is greater than 18). Rounding one number up and one number down in this situation will give an estimate closer to the exact product.

Check Your Understanding

Estimate the answer using close-but-easier numbers.

1. 644 − 292 = ?   **2.** 307 + 253 + 797 = ?   **3.** 77 × 55 = ?   **4.** 143 / 6

5. Ethan had 1,384 cards in his sports cards collection. He bought a box set with 288 cards in it. His cousin gave him another 627 sports cards for his collection. About how many cards does Ethan have in his collection now?

Check your answers in the Answer Key.

Addition Methods

Partial-Sums Addition

You can use **partial-sums addition** to find sums mentally or with paper and pencil. To use partial-sums addition, add the value of the digits in each place separately and then add the sums you just found. These are the **partial sums.**

Use an estimate to check whether your answer is reasonable.

Example

248 + 187 = **?**

Estimate the sum: 248 is close to 250, and you can round 187 to 200. 250 + 200 = 450

		2	4	8
	+	1	8	7
Add the 100s. 200 + 100 →		3	0	0
Add the 10s. 40 + 80 →		1	2	0
Add the 1s. 8 + 7 →			1	5
Add the partial sums.		4	3	5

248 + 187 = **435**

435 is reasonable because it is close to the estimate of 450.

Note To make your written work more efficient, you don't need to write the steps shown in green because steps such as estimating or adding the 1s, 10s, and 100s can be done using mental math.

You can use base-10 blocks to show partial-sums addition.

Example

Use base-10 blocks to add 248 + 187.

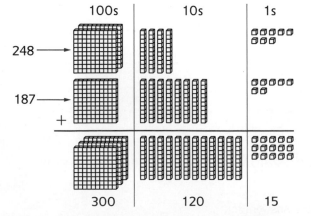

The total is 300 + 120 + 15 = 435.

248 + 187 = **435**

Note

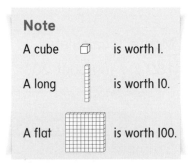

A cube ⬚ is worth 1.

A long ▯ is worth 10.

A flat ▦ is worth 100.

Column Addition

You can use **column addition** to find sums with paper and pencil.
To add numbers using column addition:

- Add the numbers in each place-value column. Write each sum in its column.

- If the sum of any column is a 2-digit number, make a trade with the column to the left.

Use an estimate to check whether your answer is reasonable.

Example

$248 + 187 = ?$

Estimate the sum: 248 is close to 250, and you can round 187 to 200.

$250 + 200 = 450$

		2	4	8
+		1	8	7
Add the numbers in each column. →		3	12	15
Trade 10 ones for 1 ten. Move the 1 ten to the tens column. →		3	13	5
Trade 10 tens for 1 hundred. Move the 1 hundred to the hundreds column. →		4	3	5

$248 + 187 = \mathbf{435}$

The answer is reasonable because 435 is close to the estimate of 450.

Check Your Understanding

Add using column addition. Estimate to check whether your answers are reasonable.

1. $\begin{array}{r} 327 \\ +252 \\ \hline \end{array}$
2. $\begin{array}{r} 67 \\ +45 \\ \hline \end{array}$
3. $\begin{array}{r} 277 \\ +144 \\ \hline \end{array}$
4. $\begin{array}{r} 2{,}268 \\ +575 \\ \hline \end{array}$
5. $\begin{array}{r} 34 \\ 54 \\ +47 \\ \hline \end{array}$

Check your answers in the Answer Key.

Number and Operations in Base Ten

U.S. Traditional Addition

U.S. traditional addition is similar to column addition. In this method, you add one column at a time, from right to left, making trades as you go. You do not record the partial sums. Use an estimate to check whether your answer is reasonable.

Example

248 + 187 = **?**

Estimate the sum: 248 is close to 250, and you can round 187 to 200.

250 + 200 = 450

Step 1 Add the 1s:

8 ones + 7 ones = 15 ones

15 ones = 1 ten + 5 ones

Write 5 in the 1s place below the line.

Write a 1 above the digits in the 10s place.

$$
\begin{array}{ccc}
 & 1 & \\
2 & 4 & 8 \\
+\ 1 & 8 & 7 \\
\hline
 & & 5 \\
\end{array}
$$

Step 2 Add the 10s:

1 ten + 4 tens + 8 tens = 13 tens

13 tens = 1 hundred and 3 tens

Write a 3 in the 10s place below the line.

Write a 1 above the digits in the 100s place.

$$
\begin{array}{ccc}
1 & 1 & \\
2 & 4 & 8 \\
+\ 1 & 8 & 7 \\
\hline
 & 3 & 5 \\
\end{array}
$$

Step 3 Add the 100s:

1 hundred + 2 hundreds + 1 hundred = 4 hundreds

Write a 4 in the 100s place below the line.

$$
\begin{array}{ccc}
1 & 1 & \\
2 & 4 & 8 \\
+\ 1 & 8 & 7 \\
\hline
4 & 3 & 5 \\
\end{array}
$$

248 + 187 = **435**

The answer is reasonable because 435 is close to the estimate of 450.

Examples

6,974 + 9,489 = **?**

Estimate the sum: 6,974 is close to 7,000, and you can round 9,489 to 9,500.

7,000 + 9,500 = 16,500

Step 1 Add the 1s:

4 ones + 9 ones = 13 ones

13 ones = 1 ten + 3 ones

Write 3 in the 1s place below the line.

Write a 1 above the digits in the 10s place.

```
      1
  6 9 7 4
+   9 4 8 9
          3
```

Step 2 Add the 10s:

1 ten + 7 tens + 8 tens = 16 tens

16 tens = 1 hundred and 6 tens

Write a 6 in the 10s place below the line.

Write a 1 above the digits in the 100s place.

```
    1 1
  6 9 7 4
+   9 4 8 9
        6 3
```

Step 3 Continue adding through the 1,000s place.

```
  1 1 1
  6 9 7 4
+ 9 4 8 9
1 6 4 6 3
```

6,974 + 9,489 = **16,463**

The answer makes sense because 16,463 is close to the estimate of 16,500.

Check Your Understanding

Add. Estimate to check whether your answers are reasonable.

1. 1,066 + 2,525 = ?

2. 9,649
   +803

3. 99,628
   +70,975

Check your answers in the Answer Key.

Subtraction Methods

Trade-First Subtraction

To use **trade-first subtraction,** compare each digit in the top number with each digit below it and make any needed trades before subtracting. To subtract numbers using trade-first subtraction:

- Look at the digits in each place and make all necessary trades.
- If each digit in the top number is greater than or equal to the digit below it, no trades are needed.
- If any digit in the top number is less than the digit below it, make a trade with the digit in the column to the left.
- Subtract separately in each column.

Use an estimate to check whether your answer is reasonable.

Example

352 − 164 = **?**

Estimate the difference: 352 is close to 350, and 164 is close to 150.

350 − 150 = 200

100s	10s	1s
3	5	2
− 1	6	4

Look at the 1s place.

2 < 4, so you need to make a trade.

100s	10s	1s
	4	12
3	5̶	2̶
− 1	6	4

So trade 1 ten for 10 ones.

Mark the problem to show the trade.

Now look at the 10s place. 4 < 6, so you need to make a trade.

100s	10s	1s
	14	
2	4̶	12
3̶	5̶	2̶
− 1	6	4
1	8	8

So trade 1 hundred for 10 tens.

Mark the problem to show the trade.

Now subtract in each column.

352 − 164 = **188**

The answer makes sense because 188 is close to the estimate of 200.

You can subtract larger numbers with 4 or more digits in the same way.

Check Your Understanding

Subtract using trade-first subtraction. Estimate to check whether your answers are reasonable.

1. 84
   −38

2. 764
   −281

3. 583
   −306

4. 3,568 − 1,086

Check your answers in the Answer Key.

Example

$352 - 164 = ?$

Estimate the difference: 352 is close to 350, and 164 is close to 150.

$350 - 150 = 200$

Use pictures of base-10 blocks to model the larger number, 352.

Write the number to be subtracted, 164, beneath the block pictures.

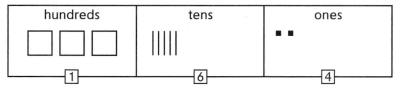

Note

A cube ■ is worth 1.

A long | is worth 10.

A flat □ is worth 100.

Think: Can I remove 1 flat from 3 flats? Yes.

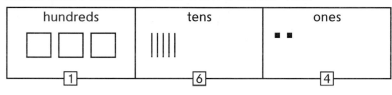

Think: Can I remove 6 longs from 5 longs? No. I need to make a trade. Trade 1 flat for 10 longs.

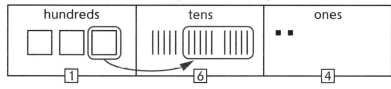

Think: Can I remove 4 cubes from 2 cubes? No. Trade 1 long for 10 cubes.

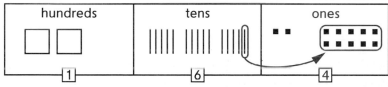

After all of the trading, the quantity of the blocks remains 352, but the blocks look like this:

hundreds	tens	ones

Now subtract the blocks that represent 164.

hundreds	tens	ones

$$\begin{array}{r} 3\ 5\ 2 \\ -\ 1\ 6\ 4 \\ \hline 1\ 8\ 8 \end{array}$$

The difference is 188.

So, $352 - 164 = \textbf{188}.$ This is a reasonable answer since the estimate was 200.

Counting-Up Subtraction

You can subtract two numbers by counting up from the smaller number to the larger number. Here is one way to subtract using **counting-up subtraction:**

- Count up to the nearest multiple of 10.

- Next count up by 10s and 100s.

- Then count up to the larger number.

Use an estimate to check whether your answer is reasonable.

Example

$325 - 38 = ?$

Estimate the difference: 38 is close to 50. 325 can stay the same since it is easy to subtract 50 from a multiple of 25.

$325 - 50 = 275$

Write the smaller number, 38.

As you count from 38 up to 325, circle each amount that you count up.

$$\begin{array}{r} 3\ \ 8 \\ +\quad ② \\ \hline 4\ \ 0 \\ +\ ⑥\ ⓪ \\ \hline 1\ \ 0\ \ 0 \\ +\ ②\ ⓪\ ⓪ \\ \hline 3\ \ 0\ \ 0 \\ +\quad ②\ ⑤ \\ \hline 3\ \ 2\ \ 5 \end{array}$$

Count up to the nearest 10.

Count up to the nearest 100.

Count up to the largest possible hundred.

Count up to the larger number.

Add the amounts you circled: $2 + 60 + 200 + 25 = 287$

You counted up 287.

$325 - 38 = \textbf{287}$

The answer makes sense because 287 is close to the estimate of 275.

Here is another way is to count up using close-but-easier numbers.

Example

$127 - 74 =$ **?**

Estimate: 127 is close to 125, and 74 is close to 75. $125 - 75 = 50$

Write the smaller number.

```
        7  4
   +      (1)      Count up to 75, a close-but-easier number.
        7  5
   +  (2  5)       Count up to 100, another close-but-easier
   1  0  0         number.
   +  (2  5)       Count up to 125, another close-but-easier
   1  2  5         number
   |     (2)       Count up to the larger number.
   1  2  7
```

Add the numbers you circled: $1 + 25 + 25 + 2 = 53$

You counted up 53.

$127 - 74 =$ **53**

This answer is reasonable because it is close to the estimate of 50.

You can use a number line to show subtraction by counting up. There are many different ways to use this method to solve subtraction problems. Here is one way:

Example

$325 - 38 = ?$

Estimate the difference: You can round 325 to 330 and round 38 to 40. $330 - 40 = 290$

Draw a number line. Mark the smaller number, 38.

38

Think: How can I get from 38 to 325?

Start at 38. As you count from 38 up to 325, circle each amount that you count up.

• Count up to the nearest 10. Mark 40 on the number line.

• Count up to the nearest 100. Mark 100.

• Count up to the largest possible hundred. Mark 300.

• Count up to the larger number. Mark 325.

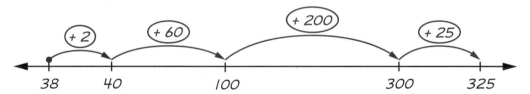

Add the amounts you circled: $2 + 60 + 200 + 25 = 287$

You counted up by 287.

$325 - 38 = \textbf{287}$

The answer makes sense because 287 is close to the estimate of 290.

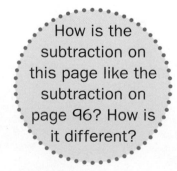

How is the subtraction on this page like the subtraction on page 96? How is it different?

Here is another way:

Example

$325 - 38 = ?$

Draw a number line. Mark the smaller number, 38.

38

Think: How can I get from 38 to 325?

Start at 38. As you count from 38 up to 325, circle each amount that you count up.

- Count up by hundreds from 38. Mark each point on the number line.
- Count up by a multiple of 10. Count up 60. Mark 298.
- Count up to the larger number. Mark each point up to 325.

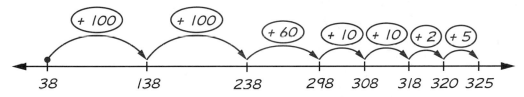

Add the amounts you circled: $100 + 100 + 60 + 10 + 10 + 2 + 5 = 287$

You counted up 287.

$325 - 38 = \mathbf{287}$

The answer makes sense because 287 is close to the estimate of 290.

Check Your Understanding

Subtract using counting-up subtraction. Estimate to check whether your answers are reasonable.

1. $172 - 74$ **2.** $403 - 157$ **3.** $468 - 253$ **4.** $915 - 56$

Check your answers in the Answer Key.

U.S. Traditional Subtraction

You can subtract numbers using **U.S. traditional subtraction.** Start at the right. Subtract column by column and make any necessary trades. Use an estimate to check whether your answer is reasonable.

Example

572 − 385 = **?**

Estimate the difference: You can round 572 to 600 and 385 to 400.

600 − 400 = 200

Step 1 Start with the 1s:

Since 2 < 5, you need to make a trade.

Trade 1 ten for 10 ones:

572 = 5 hundreds + 6 tens + 12 ones.

Subtract the 1s: 12 ones − 5 ones = 7 ones.

$$\begin{array}{r} 5 \ \overset{6}{\cancel{7}} \ \overset{12}{\cancel{2}} \\ - \ 3 \ 8 \ 5 \\ \hline 7 \end{array}$$

Step 2 Go to the 10s:

Since 6 < 8, you need to make a trade.

Trade 1 hundred for 10 tens:

572 = 4 hundreds + 16 tens + 12 ones.

Subtract the 10s: 16 tens − 8 tens = 8 tens.

$$\begin{array}{r} \overset{4}{\cancel{5}} \ \overset{\overset{16}{\cancel{6}}}{\cancel{7}} \ \overset{12}{\cancel{2}} \\ - \ 3 \ 8 \ 5 \\ \hline 8 \ 7 \end{array}$$

Step 3 Go to the 100s:

Subtract the 100s: 4 hundreds − 3 hundreds = 1 hundred.

$$\begin{array}{r} \overset{4}{\cancel{5}} \ \overset{\overset{16}{\cancel{6}}}{\cancel{7}} \ \overset{12}{\cancel{2}} \\ - \ 3 \ 8 \ 5 \\ \hline 1 \ 8 \ 7 \end{array}$$

572 − 385 = **187**

The answer makes sense because 187 is close to the estimate of 200.

Example

$802 - 273 = ?$

Estimate the difference: 802 is close to 800, and you can round 273 to 300.

$800 - 300 = 500$

Step 1 Start with the 1s:

Since 2 < 3, you need to make a trade.

There are no tens in the tens place in 802, so trade 1 hundred for 10 tens and then trade 1 ten for 10 ones:

802 = 7 hundreds + 9 tens + 12 ones.

Subtract the 1s: 12 ones − 3 ones = 9 ones.

```
        9
   7   10   12
   8    0    2
 −  2   7    3
 ─────────────
                9
```

Step 2 Go to the 10s:

Since 9 > 7, you do not need to make a trade.

Subtract the 10s: 9 tens − 7 tens = 2 tens.

```
        9
   7   10   12
   8    0    2
 −  2   7    3
 ─────────────
           2    9
```

Step 3 Go to the 100s:

Subtract the 100s: 7 hundreds − 2 hundreds = 5 hundreds.

```
        9
   7   10   12
   8    0    2
 −  2   7    3
 ─────────────
      5    2    9
```

$802 - 273 = \textbf{529}$

The answer makes sense because 529 is close to the estimate of 500.

Check Your Understanding

Subtract using U.S. traditional subtraction. Estimate to check whether your answers are reasonable.

1. $441 - 386 = ?$

2.
$$\begin{array}{r} 308 \\ - 278 \\ \hline \end{array}$$

3.
$$\begin{array}{r} 8,694 \\ - 2,708 \\ \hline \end{array}$$

Check your answers in the Answer Key.

Extended Multiplication Facts

You are using **extended multiplication facts** when you combine basic multiplication facts and multiplying with multiples of 10.

You can use patterns to find shortcuts for multiplying whole numbers by 10s, 100s, and 1,000s.

Note Strategies for the basic multiplication facts can be found on pages 42–44.

Examples

Notice the pattern in the number of zeros in the product when a whole number is multiplied by 10, 100, and 1,000.

10 * 8 = 80	10 * 57 = 570	10 * 40 = 400
100 * 8 = 800	100 * 57 = 5,700	100 * 40 = 4,000
1,000 * 8 = 8,000	1,000 * 57 = 57,000	1,000 * 40 = 40,000

You can use mental math to find the products of basic facts and multiples of 10.

Examples

8 * 60 = **?**

Think: 8 [6s] = 48
Then 8 [60s] is 10 times as much because 60 is 6 tens.

8 * 60 = 10 * 48 = 480
So, 8 * 60 = **480.**

4,000 * 5 = **?**

Think: 4 [5s] = 20
Then 4,000 [5s] is 1,000 times as much because 4,000 is 4 thousands.

4,000 * 5 = 1,000 * 20 = 20,000
So, 4,000 * 5 = **20,000.**

You can use a similar method to find products when both factors are multiples of 10.

Examples

20 * 50 = **?**

Think: 2 [50s] = 100
Then 20 [50s] is 10 times as much.

20 * 50 = 10 * 100 = 1,000
So, 20 * 50 = **1,000.**

200 * 90 = **?**

Think: 2 [90s] = 180
Then 200 [90s] is 100 times as much.

200 * 90 = 100 * 180 = 18,000
So, 200 * 90 = **18,000.**

Check Your Understanding

Solve these problems mentally.

1. 8 * 100   **2.** 1,000 * 41   **3.** 6 * 500

4. 4,000 * 9   **5.** 60 * 40   **6.** 500 * 50

Check your answers in the Answer Key.

Area Models for Multiplication

An **area model** is a way to represent multiplication problems. The length and width of a rectangle represent the **factors,** and the area of the rectangle represents the **product.**

Alex wants to create a flower garden that has 7 rows with 12 flowers in each row. How many flowers does Alex need to plant?

You can draw a rectangular array to solve the problem:

$7 * 12 = 84$ flowers

You can draw an area model to solve the problem:

12

7 | $7 * 12 = 84$

$7 * 12 = 84$ flowers

Alex needs to plant 84 flowers.

Number and Operations in Base Ten

When multiplying larger factors, it may take too long to draw an array. Area models can be drawn easily and are a more efficient way to represent multiplication problems.

Area models can be divided, or **partitioned,** into smaller rectangles. The length or width (or both) can be decomposed into numbers that are easier to work with. You can create an equation to find the area of the whole rectangle by adding the areas of the smaller rectangles. There are different ways to partition a rectangle. The examples below show two ways.

Examples

$7 * 12 = ?$

Estimate: $7 * 10 = 70$. Since 12 was rounded down, the product will be more than 70.

One way:

Partition the length of 12 into lengths of 10 and 2. This creates two smaller rectangles.

Multiply to find the areas of the two smaller rectangles.

Add the areas of the two smaller rectangles to find the total area of the largest rectangle.

$7 * 12 = $ **84**

This is reasonable since 84 is more than the estimate of 70.

Another way:

Partition the length of 12 into lengths of 10 and 2.

Partition the width of 7 into widths of 5 and 2. This creates four smaller rectangles.

Multiply to find the areas of the four smaller rectangles.

Add the areas of the four smaller rectangles to find the total area of the largest rectangle.

$7 * 12 = $ **84**

This is reasonable since 84 is more than the estimate of 70.

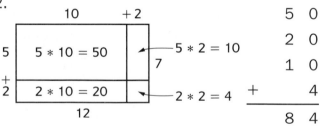

Partitioning rectangles is especially helpful when either of the factors are larger numbers.

Example

$4 * 68 = ?$

Estimate: $4 * 70 = 280$

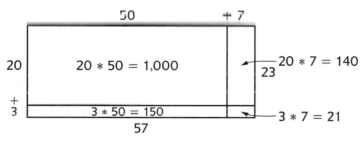

$$
\begin{array}{r}
2\ 4\ 0 \\
+\quad\ \ 3\ 2 \\
\hline
2\ 7\ 2
\end{array}
$$

$4 * 68 = \mathbf{272}$

This answer is reasonable since it is close to the estimate of 280.

You can use area models to solve multiplication problems when both factors are multidigit numbers.

Example

$23 * 57 = ?$

Estimate: $20 * 60 = 1,200$

	50	+ 7
20	$20 * 50 = 1,000$	$20 * 7 = 140$
+ 3	$3 * 50 = 150$	$3 * 7 = 21$

57   23

$$
\begin{array}{r}
1,\ 0\ 0\ 0 \\
1\ 5\ 0 \\
1\ 4\ 0 \\
+\quad\ \ 2\ 1 \\
\hline
1,\ 3\ 1\ 1
\end{array}
$$

$23 * 57 = \mathbf{1,311}$

This is reasonable since it is close to the estimate of 1,200.

Multiplication Methods

Partial-Products Multiplication

In **partial-products multiplication,** the value of each digit in one factor is multiplied by the value of each digit in the other factor. The sum of the **partial products** is the final product, or the answer to the original problem.

Partial-products multiplication is similar to partitioning rectangles in area models, without having to draw a rectangle.

When partitioning a rectangle:

- Factors are often broken into place-value parts and the partial products are recorded inside the smaller rectangles. Each partial product represents the area of the smaller rectangle.

- The areas of the smaller rectangles are added together to find the total area of the rectangle.

When using partial-products multiplication:

- Factors are mentally broken into place-value parts, and the partial products are recorded beneath the problem.

- The partial products are added together to find the total product.

Example

$5 * 26 = ?$

Multiply using partial-products multiplication.

Estimate the product: 26 is close to 25.

$5 * 25 = 125$ since 4 [25s] = 100 and 5 [25s] is 25 more, or 125.

Think of 26 as 20 + 6.

Multiply each part by 5.

```
                2   6
        *           5
        ————————————
5 * 20 →    1   0   0
5 * 6  →  +     3   0
        ————————————
            1   3   0
```

Add the partial products.

$5 * 26 = $ **130**

The answer makes sense because 130 is close to the estimate of 125.

Multiply by partitioning using an area model.

```
        20        + 6      ⌐ 5 * 6 = 30
    ┌─────────────┬────┐     100
  5 │ 5 * 20 = 100 │    │   + 30
    └─────────────┴────┘     130
          └──── 26 ────┘
```

Partial-Products Multiplication with Multidigit Factors

When you use partial-products multiplication with multidigit factors, it is important that all of the partial products are included. Use an estimate to check whether your answer is reasonable.

Example

$34 * 26 = ?$

Estimate the product: 34 rounds down to 30, and 26 rounds up to 30. $30 * 30 = 900$

Think of 34 as 30 + 4.				3	4
Think of 26 as 20 + 6.		$*$		2	6
Multiply each part of 34	$20 * 30 \rightarrow$		6	0	0
by each part of 26.	$20 * 4 \rightarrow$			8	0
	$6 * 30 \rightarrow$		1	8	0
	$6 * 4 \rightarrow$	$+$		2	4
Add the four partial products.			8	8	4

$34 * 26 = \mathbf{884}$

The answer makes sense because 884 is close to the estimate of 900.

Note If you can estimate and multiply the partial products using mental math, then you do not need to write the steps shown in green.

Check Your Understanding

Multiply using partial-products multiplication. Estimate to check whether your answers are reasonable.

1. $73 * 5$ **2.** $43 * 63$ **3.** $40 * 27$ **4.** $22 * 22$ **5.** $316 * 3$

Check your answers in the Answer Key.

Lattice Multiplication

Lattice multiplication has been used for hundreds of years. It is based on placing answers to basic multiplication facts in boxes, and then adding along diagonals. The box with cells and diagonals is called a **lattice.**

Lattice multiplication works because each diagonal is the same as a place-value column. The lattice is like a place-value chart. The far right-hand diagonal is the ones place, the next diagonal to the left is the tens place, the third diagonal is the hundreds place, and so on.

Example

3 * 45 = **?**

Estimate the product: 45 is between 40 and 50, so the product should be between 3 * 40 and 3 * 50, or between 120 and 150.

Write 45 above the lattice.
Write 3 on the right side of the lattice.
Multiply 3 * 5. Then multiply 3 * 4.
Write the answers in the lattice as shown.

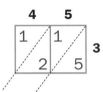

Add the numbers along each diagonal, starting at the right.
Read the answer. 3 * 45 = **135**

hundreds tens ones

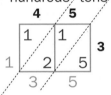

135 is reasonable since it falls between 120 and 150.

Example

34 * 26 = **?**

Estimate the product: 34 rounds down to 30, and 26 rounds up to 30.
30 * 30 = 900

Write the 26 above the lattice and 34 on the right side of the lattice.

Multiply 3 * 6. Then multiply 3 * 2.
Multiply 4 * 6. Then multiply 4 * 2.
Write the answers in the lattice as shown.

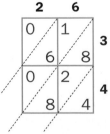

Add the numbers along each diagonal, starting at the right.

When the numbers along a diagonal add to 10 or more:
- record the ones digit, then
- add the tens digit to the sum along the diagonal above.

Read the answer. 34 * 26 = **884**

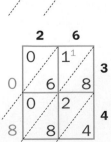

The answer makes sense because it is close to the estimate of 900.

Basic Division Facts

A division fact can represent sharing equally or forming equal groups.

Sharing equally:

$35 / 5 = ?$ 5 people share 35 pennies

How many pennies does each person get?

$35 / 5 = 7$ Each person gets 7 pennies.

Forming equal groups:

$35 / 5 = ?$ There are 35 apples in all.

5 apples are put into each bag.

How many bags can be filled?

$35 / 5 = 7$ 7 bags can be filled.

If you don't remember a basic fact, try one of the following methods:

Use Counters or Draw a Picture

To find $35 / 5$, start with 35 objects.

Think: How many 5s in 35?

Make groups or circle groups of 5 objects each. Count the groups.

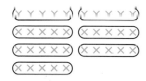

Skip Count Down

To find $35 / 5$, start at 35 and count by 5s down to 0. Use your fingers to keep track as you skip count.

35, 30, 25, 20, 15, 10, 5, 0
That's 7 skips.

Use Known Multiplication Facts

Think about fact families. Every division fact is related to a multiplication fact. For example, if you know that $5 * 7 = 35$ or $7 * 5 = 35$, you also know that $35 / 5 = 7$ and $35 / 7 = 5$.

Note You can multiply by 0, but you cannot divide by 0. For example, $0 * 9 = 0$, $9 * 0 = 0$, and $0 / 9 = 0$, but $9 / 0$ has no answer. Consider that to find $8 / 2 = ?$, you can think about the multiplication fact $2 * ? = 8$. Since $2 * 4 = 8$, you know that $8 / 2 = 4$. To find an answer for $9 / 0 = ?$, think about the multiplication fact $0 * ? = 9$. There is no number that will work for **?** to make the number sentence true, so $9 / 0$ has no answer.

University of Chicago

Extended Division Facts

You are using **extended division facts** when you combine basic division facts and multiples of 10.

One way to find answers to extended division facts is to relate the problem to basic facts and to multiplication with 10, 100, and 1,000.

Examples

240 / 3 = **?**

Think: 24 / 3 = 8

Because 240 is 24 [10s],
240 / 3 is 10 times as much
as 24 / 3 = 8.

240 / 3 = 10 * 8 = 80

So, 240 / 3 = **80.**

15,000 / 5 = **?**

Think: 15 / 5 = 3

Because 15,000 is 15 [1,000s],
15,000 / 3 is 1,000 times as
much as 15 / 5 = 3.

15,000 / 5 = 1,000 * 3 = 3,000

So, 15,000 / 5 = **3,000.**

You can also use the relationship between multiplication and division to solve extended division facts, especially when the numbers are large.

Example

1,800 / 30 = **?**

Write 1,800 / 30 as a missing factor
multiplication number sentence.

What is the related basic division fact?

Try 6 as the missing factor for your
multiplication number sentence.

Try 6 * 10, or 60, as your missing factor.

1,800 = ? * 30

18 / 3 = 6

6 * 30 = 180 Too small.
You want 1,800. So you need a number
that is 10 times as much as 6.

60 * 30 = 1,800
This works.

Since 60 * 30 = 1,800, then 1,800 / 30 = 60.

1,800 / 30 = **60**

Check Your Understanding

Solve these problems mentally.

1. 4,000 / 100   **2.** 47,000 / 1,000   **3.** 28,000 / 7   **4.** 4,800 / 10   **5.** 4,800 / 80

Check your answers in the Answer Key.

Division Methods

Different symbols may be used to indicate division. For example, "94 divided by 6" may be written as $94 \div 6$, $6\overline{)94}$, $94 / 6$, or $\frac{94}{6}$.

- The number being divided is the **dividend.**

- The number that divides the dividend is the **divisor.**

- The answer to a division problem is the **quotient.**

- Some numbers cannot be divided evenly. When this happens, the answer includes a quotient and a **remainder.**

Four ways to show "123 divided by 4"
$123 \div 4 \rightarrow 30 \text{ R3}$ $123 / 4 \rightarrow 30 \text{ R3}$
$\begin{array}{r} 30 \text{ R3} \\ 4\overline{)123} \end{array}$ $\frac{123}{4} \rightarrow 30 \text{ R3}$
123 is the dividend.
4 is the divisor.
30 is the quotient.
3 is the remainder.

Array and Area Models for Division

An **array** is helpful for representing division problems. The total number of items in an array represents the dividend. When you use an array to solve a division problem, you are finding either the number of rows or the number of items in a row.

Example

Jada is helping her father lay tile on a rectangular portion of their kitchen floor. 70 square-foot tiles fill this portion of the floor. 5 of the tiles fit along the length of the space. How many rows are needed to fill the space?

Draw an array to solve the problem:

Think: How many rows of 5 tiles are in 70 tiles?

Write a number model with a letter standing for the unknown number of rows: $70 / 5 = r$.

- 1 row has 5 square tiles.
- Each row repeats until you reach 70 tiles in all.
- Skip count by 5 until you reach 70:
  5, 10, 15, 20, 25, 30, 35, 40, 45, 50, 55, 60, 65, 70.
- The number of times you skip counted by 5 tells you the number of rows.

70 tiles / 5 tiles in each row = 14 rows

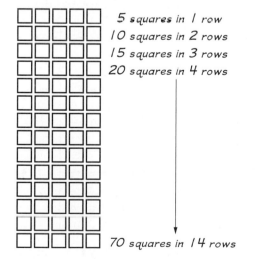

5 squares in 1 row
10 squares in 2 rows
15 squares in 3 rows
20 squares in 4 rows

70 squares in 14 rows

Jada needs 14 rows of 5 tiles to fit all 70 tiles.

You have used **area models** to represent multiplication problems. You can also use an area model to help solve a division problem. The total area of the rectangle represents the **dividend,** and one of the side lengths represents the **divisor.** The unknown side length is the **quotient**.

Example

Jada and her father are laying tile on a section of floor that is 5 feet wide and has an area of 70 square feet. Use an area model to find the the length of the floor.

First, draw a rectangle and write a number model to represent the number story. You can use a variable to stand for the unknown length:

$70 / 5 = r$

5 ft

70 sq ft   r

Second, use basic facts to find a group of 5 that gets you closer to 70.

$10 * 5 = 50$ represents a 10 ft by 5 ft section. Partition the rectangle to represent the part that is 50 sq ft.

20 more are needed to make 70 sq ft.

5 ft

$10 * 5 = 50$ sq ft   10 ft

$\begin{array}{r} 70 \\ -\,50 \\ \hline 20 \end{array}$

Area = 70 sq ft

Third, find the length of the missing part.

$4 * 5 = 20$, so a 4 ft by 5 ft section makes up the remaining area. The length of the missing part is 4 ft.

5 ft

$10 * 5 = 50$ sq ft   10 ft

$4 * 5 = 20$ sq ft   4 ft

$\begin{array}{r} 70 \\ -\,50 \\ \hline 20 \\ -\,20 \\ \hline 0 \end{array}$

Area = 70 sq ft

Fourth, find the total length by adding the lengths of each partition.

The total length is $10 + 4 = 14$ ft.

5 ft

$10 * 5 = 50$ sq ft   10 ft

$4 * 5 = 20$ sq ft   4 ft

$\begin{array}{r} 70 \\ -\,50 \\ \hline 20 \\ -\,20 \\ \hline 0 \end{array}$

+

14 ft

Area = 70 sq ft

$70 / 5 = 14$

The length of the floor is 14 feet.

Partial-Quotients Division

Partial-quotients division is similar to partitioning rectangles in the area model, without having to draw a rectangle. At each step in partial-quotients division, you find a partial answer (called a **partial quotient**). You add the partial answers to find the quotient.

Study this example. To find the number of [5s] in 70, first find partial quotients, then add them. Record the partial quotients in a column to the right of the original problem.

Example

$70 / 5 = ?$

Estimate: $50 / 5 = 10$ and $100 / 5 = 20$, so the quotient will be between 10 and 20.

Write the partial quotients in this column.

```
5) 70
 – 50   10    Think: How many [5s] are in 70? At least 10.
 ─────         The first partial quotient is 10. 10 * 5 = 50 Subtract 50 from 70.
   20
 – 20    4    Think: How many [5s] are in 20? There are 4.
 ─────         The second partial quotient is 4. 4 * 5 = 20 Subtract 20 from 20.
    0   14    Add the partial quotients.
```

Quotient

The answer is **14.** Record the answer as $5\overline{)70}$ with 14 above, or rewrite $70 / 5 = 14$.

The answer makes sense because 14 is between 10 and 20, which matches the estimate.

There are different ways to find partial quotients when you use partial-quotients division.

Example

$228 / 6 = ?$

Estimate: 228 is close to 240. $240 / 6 = 40$

One way:
```
6)228
– 180  30
─────
   48
–  30   5
─────
   18
–  18   3
─────
    0  38
```

Another way:
```
6)228
– 120  20
─────
  108
–  60  10
─────
   48
–  48   8
─────
    0  38
```

Still another way:
```
6)228
– 180  30
─────
   48
–  48   8
─────
       38
```

The quotient, **38,** is the same each way.

The answer is reasonable since it is close to the estimate of 40.

One way of doing partial-quotients division is to ask yourself a series of "at least/not more than" questions using extended multiplication facts of the divisor. Study the example below.

Example

2,268 / 7 = **?**

Estimate: 2,268 is close to 2,100. 2,100 / 7 = 300

Find the number of [7s] in 2,268 using extended multiplication facts.
Are there at least **100** [7s] in 2,268?
Yes. **100** * 7 = 700 and 700 < 2,268.
Are there at least **200** [7s] in 2,268?
Yes. **200** * 7 = 1,400 and 1,400 < 2,268.
Are there at least **300** [7s] in 2,268?
Yes. **300** * 7 = 2,100 and 2,100 < 2,268.
Are there at least **400** [7s] in 2,268?
No. **400** * 7 = 2,800 and 2,800 > 2,268.
So there are at least **300** [7s] in 2,268.

168 remains. Find the number of [7s] in 168 using extended multiplication facts.
Are there at least **100** [7s] in 168?
No. **100** * 7 = 700 and 700 > 168.
Are there at least **50** [7s] in 168?
No. **50** * 7 = 350 and 350 > 168.
Are there at least **10** [7s] in 168?
Yes. **10** * 7 = 70 and 70 < 168.
Are there at least **20** [7s] in 168?
Yes. **20** * 7 = 140 and 140 < 168.
Are there at least **30** [7s] in 168?
No. **30** * 7 = 210 and 210 > 168.
So there are at least **20** [7s] in 168.
28 remains. There are **4** [7s] in 28.
Add the partial quotients: 300 + 20 + 4 = 324

```
7)2,268
- 2,100 | 300
   168
-  140  |  20
    28
-   28  |   4
     0    324
```

2,268 / 7 = **324**

This answer is reasonable since it is close to the estimate of 300.

Check Your Understanding

Solve. Ask yourself "at least/not more than" questions using extended multiplication facts of the divisor for each division problem. Estimate to check whether your answers are reasonable.

1. 384 / 3   **2.** 7)441   **3.** 2,104 ÷ 8

Check your answers in the Answer Key.

Division with Remainders

When numbers cannot be divided evenly, the answer includes a quotient and a **remainder.** The way you represent the quotient and remainder depends on the problem situation.

There are several ways to think about remainders:

- Ignore the remainder. Use the quotient as the answer.
- Round the quotient up to the next whole number.
- Rewrite the remainder as a fraction or decimal. Use this fraction or decimal as part of the answer.

Example

Suppose 3 people share 14 counters equally. How many counters will each person get?

A number model is 14 / 3.

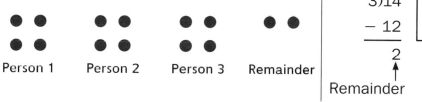

Person 1   Person 2   Person 3   Remainder

$$3\overline{)14}$$
$$-\ 12 \quad | \quad 4$$
$$\overline{\quad 2 \quad | \quad 4}$$

Remainder   Quotient

14 / 3 → 4 R2

Ignore the remainder. You cannot split the remaining counters among 3 people.

Answer: Each person will get 4 counters, with 2 counters left over.

Example

Suppose 14 photos are placed in a photo album. How many pages are needed if 3 photos can fit on a page?

A number model is 14 / 3.

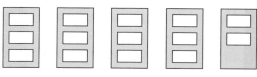

Page 1   Page 2   Page 3   Page 4   Page 5

14 / 3 → 4 R2

Round the quotient up to the next whole number. A fifth page is needed to include all the photos. The album will have 4 pages filled and another page only partially filled.

Answer: 5 pages are needed.

Calculators

Beginning in the 1600s, calculating machines that used levers, wheels, and gears were invented.

By the late 19th and early 20th century, hand-cranked pinwheel calculators were in widespread use. These calculators used wheels with interlocking gears to carry out calculations. Turning the wheels in one direction performed addition while turning them in the opposite direction carried out subtraction.

Electric-powered mechanical calculators were invented in the 1960s and quickly began replacing the older hand-powered models. These calculators used two keyboards to calculate. Because of their speed and versatility, electric-powered mechanical calculators were used extensively through the 1970s.

Pocket-size calculators first became available in the 1970s after the invention of the microprocessor. Modern calculators like this one can perform a wide variety of mathematical functions and can even do them faster than the most powerful computers of 40 years ago. As their prices have dropped, these portable devices have seen a rapid growth in popularity.

Computer Firsts

The first all-purpose electronic computer, called the Electrical Numerical Integrator and Calculator (ENIAC), was built in 1946.

Compared with today's personal computers and handheld calculators, ENIAC was huge. It weighed 30 tons, used about 18,000 vacuum tubes, and took up an entire room. When it was built, ENIAC could compute about 1,000 times faster than any previous device.

UNIVAC, which stands for Universal Automatic Computer, became the world's first commercially available electronic computer in 1951. "Universal" meant it could solve problems for scientists, engineers, and businesses. "Automatic" meant the instructions in the computer could be stored in the computer's memory.

Along with computers came "computer bugs." The first actual computer bug was found by Grace Hopper on September 9, 1947.

Commodore Grace M. Hopper, United States Naval Reserve, 1985.

While testing the Mark II Aiken Relay Calculator, Grace Hopper encountered a problem: a moth was blocking one of the computer's relays. She removed the moth and taped it to her notes with the entry, "First actual case of bug being found." She later explained that she had "debugged" the machine.

Personal Computers

By the 1970s, personal computers were invented. The personal computer was made for a single person to use. It fit on a desk and was much less expensive than previous computers.

The Apple II computer was introduced in 1977. The computer could use a standard television set for its monitor.

The IBM Personal Computer (PC) was first sold in 1981. It came fully assembled and included hardware and software made by other companies. Within 10 years, there were 50 million computers modeled after the PC.

The invention of laptop computers changed people's lives by making it possible to take a computer almost anywhere.

Today's personal computers are very different from the first ones invented. They often have faster processors, more memory, and a variety of other features that make them convenient and easy to use. The Internet, a system of connected networks from all around the world, began in the 1980s and gained popularity in the 1990s. Now almost all computers can be connected to the Internet, which people use to do research, share documents, send email, stream music and videos, and for many other purposes.

Physicist Tim Berners-Lee invented the World Wide Web, a collection of linked websites and resources, in 1989. The first websites on the Internet became available to the public in 1991.

Apple's iMac computer has advanced to using a thin widescreen LCD, or liquid-crystal display.

Students now use the World Wide Web in all areas of their lives. The Internet provides access to news, research, people, and entertainment all around the world.

Tablets can perform many of the same functions as laptop computers, but they are smaller and more portable. They have a touch screen that can be used instead of a keyboard and mouse.

Fractions

Fractions were invented thousands of years ago to name numbers that are between **whole numbers.** People likely used these in-between numbers for making more precise measurements. The word **fraction** comes from the Latin word *frangere,* which means "to break." A common use for fractions is to name parts of wholes, where the **whole** is "broken" into equal-size pieces. Fractions are also used to represent other mathematical ideas such as division and ratios.

Example

Two parts of the garden have flowers. The rest of the garden has vegetables. How much of the garden has vegetables?

the whole garden

the whole garden divided into 5 equal parts

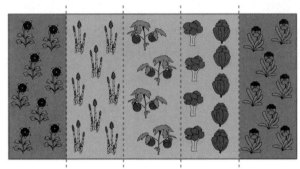

Less than the whole garden has vegetables. A fraction is needed to name the amount of the garden that has vegetables.

The garden is divided into 5 equal parts. The equal-size parts are named *fifths.* Three of the parts have vegetables. Here are ways you can describe the parts of the garden with vegetables:

3 out of 5 equal parts   three-fifths   3-fifths $\frac{3}{5}$

Example

You can use fractions to describe other features of the garden.

Two of the five equal-size parts have flowers. The parts of the garden with flowers can be named:

2 out of 5 equal parts   two-fifths   2-fifths $\frac{2}{5}$

The whole garden, or all 5 of the equal-size parts, can be named:

5 out of 5 equal parts   five-fifths   5-fifths $\frac{5}{5}$ 1 whole

---

Reading and Writing Fractions

You can write a fraction in different ways. For example, one-half can be written as 1 out of 2 equal parts, 1 half, or $\frac{1}{2}$.

The numbers $\frac{7}{8}$, $\frac{3}{4}$, $\frac{5}{4}$, $\frac{4}{4}$, and $\frac{25}{100}$ are all fractions written in fraction notation. Together, the numerator above the fraction bar and the denominator below the fraction bar describe the amount of the whole that a fraction represents. The **denominator** describes how many equal parts it takes to make the whole, which determines the size of each part. The denominator cannot be 0. The **numerator** describes the number of equal-size parts that are being considered.

When reading a fraction, say the numerator first. Then say the size of the equal parts represented by the denominator. For this fraction, you say "three-fourths."

numerator $\longrightarrow$ $\frac{3}{4}$ three-fourths
denominator $\longrightarrow$

Did You Know?

Arab mathematicians began to use the horizontal fraction bar around the year 1200. They were the first to write fractions as we do today.

Note In *Fourth Grade Everyday Mathematics,* the whole for a fraction is often shown using a "whole box." In the example below, the whole is the entire pizza. The whole box shows the whole is a circle to represent the pizza.

Example

Write a fraction for the amount of the pizza that remains on the plate.

The whole pizza is in the shape of a circle.

The pizza is divided into 8 equal parts. So the size of each part is an eighth. The denominator is 8.

Count the pieces of pizza remaining:
1 eighth, **2** eighths, **3** eighths, **4** eighths, and **5** eighths. In all, 5 of the 8 pieces remain. The numerator is 5.

Whole

circle

Write the fraction:

$\frac{5}{8}$ ←——— The **numerator** 5 tells the number of parts remaining.
←——— The **denominator** 8 tells the number of equal parts in the **whole.** Since there are 8 equal parts, the size of each part is an eighth of the whole.

$\frac{5}{8}$ of the pizza remains on the plate. Other names for this fraction are five-eighths and 5 out of 8 equal parts.

Example

Write a fraction for the part of the collection of cars that is red.

The whole is the collection of cars. There are 4 cars in the collection. Each car is a fourth of the collection. The denominator is 4. 1 of the cars is red, so the numerator is 1.

Whole
4 cars

You can say, "One-fourth of the collection of cars is red" or "1 out of 4 cars is red."

You can write $\frac{1}{4}$ to represent the part of the collection that is red.

To draw a picture of a fraction, use the denominator to figure out the size of the parts. You may choose the whole. Draw the whole with the number of equal parts given by the denominator. Use the numerator to count the number of parts to shade.

Example

Show $\frac{4}{5}$ of a whole.

Draw a whole. A long rectangle can be the whole.

The denominator tells how many equal parts to show in the whole. The denominator is 5. The size of the equal parts is fifths. Divide the rectangle into 5 equal parts.

The numerator tells how many parts to count and shade. The numerator is 4. The number of parts to count and shade is 4.

This picture shows $\frac{4}{5}$ of a whole.

$\frac{4}{5}$ of the rectangle is shaded.

The amount a fraction represents depends on the size of the whole.

Example

How does $\frac{1}{2}$ of the blue ribbon compare to $\frac{1}{2}$ of the green ribbon?

$\frac{1}{2}$ of the blue ribbon is shorter than $\frac{1}{2}$ of the green ribbon because the whole blue ribbon is shorter than the whole green ribbon.

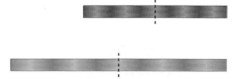

Mark Steinmetz

Fractions Equal To and Greater Than One

Some fractions represent numbers equal to one or greater than one. Just as in other fractions, the denominator describes the size of the parts by telling the number of parts into which the whole is partitioned, or divided. The numerator describes the number of parts being considered. If the numerator is equal to the denominator, the amount represented is equal to one whole and the fraction is equal to 1.

> **Note** Fractions greater than one are sometimes called *improper fractions,* although there is nothing wrong with them. It is often easier to use fractions written with the numerator greater than the denominator when using number sentences to solve problems.

Example

$\frac{4}{4}$ means 4 out of 4 equal parts. That's all of the parts of 1 whole.

$\frac{4}{4}$ = 1 whole

$\frac{6}{6}$ means 6 out of 6 equal parts. That's all of the parts of 1 whole.

$\frac{6}{6}$ = 1 whole

When the numerator is larger than the denominator, the amount represented is greater than one whole and the fraction is greater than 1.

Example

What does $\frac{13}{5}$ (thirteen-fifths) mean?

The numerator is greater than the denominator. You can show $\frac{13}{5}$ this way: So the fraction represents an amount that is greater than one.

$\frac{13}{5}$ ⟵ The *numerator* 13 tells the number of parts.

⟵ The *denominator* 5 tells the size of each part. Since there are 5 equal parts in a whole, the parts are fifths.

Fractions greater than 1 can be written as **mixed numbers.** A mixed number has a whole-number part and a fraction part. In the mixed number $2\frac{3}{5}$, the whole-number part is 2 and the fraction part is $\frac{3}{5}$. A mixed number is equal to the sum of the whole-number part and the fraction part: $2\frac{3}{5} = 2 + \frac{3}{5}$.

Example

Write $\frac{13}{5}$ as a mixed number.

$\frac{5}{5}$ = 1 whole. Show $\frac{13}{5}$ as two whole circles and three-fifths of another circle.

$\frac{13}{5}$ can be written as $2\frac{3}{5}$.

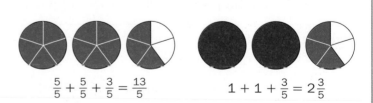

$\frac{5}{5} + \frac{5}{5} + \frac{3}{5} = \frac{13}{5}$ $1 + 1 + \frac{3}{5} = 2\frac{3}{5}$

Meanings of Fractions

Fractions can have different meanings.

Parts of Regions

Fractions can be used to name part of a whole region.

The circle is the whole.

$\frac{5}{8}$ of the circle is shaded.

$\frac{3}{8}$ of the circle is not shaded.

The rectangle is the whole.

$\frac{1}{3}$ of the rectangle is shaded.

$\frac{2}{3}$ of the rectangle is not shaded.

Parts of Collections

Fractions can be used to name part of a whole collection of objects.

The whole is all of the shapes in this collection.

$\frac{3}{10}$ of this collection of shapes are yellow.

$\frac{2}{10}$ of this collection of shapes are triangles.

Points on a Number Line

Fractions can be used to mean distance from 0 on a number line. The whole is the distance from 0 to 1, and each whole can be divided into equal distances.

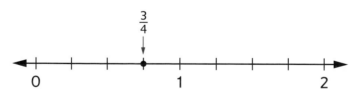

The point on the number line is located at $\frac{3}{4}$.

Division

Fractions can be used to mean division. The numerator is divided by the denominator.

The division problem 3 divided by 4 can be written in any of these ways:

$3 \div 4 \qquad 4\overline{)3} \qquad 3 / 4 \qquad \frac{3}{4}$

Uses of Fractions

Fractions have many uses in everyday life.

Rulers	Fractions can be used to name lengths measured with rulers, metersticks, yardsticks, and tape measures. The ribbon is $1\frac{3}{4}$ inches long.	
Signs for Distance	Fractions can be used to tell how far it is to a destination. A scenic view is $\frac{3}{4}$ mile ahead.	
Measuring Spoons	Fractions can be used when measuring ingredients with a measuring spoon. The amount in the measuring spoon measures $\frac{1}{2}$ teaspoon.	
Measuring Cups	Fractions can be used when measuring ingredients with a measuring cup. The liquid in the measuring cup measures $2\frac{1}{2}$ cups.	
Money	Fractions can be used to name the value of coins. A dollar is equal to 100 cents. A quarter is worth 25 cents. So this coin is worth 25 out of 100 cents, or $\frac{25}{100}$ of a dollar. $\frac{25}{100}$ is the same as $\frac{1}{4}$ of a dollar.	1 quarter 1 dollar This coin is called a *quarter*. It is worth $\frac{1}{4}$, or one-quarter, of a dollar.

Fraction Circles

There are many ways to represent fractions. One way is with fraction circles.

In the set of fraction circles used with *Everyday Mathematics*, the red circle is the largest fraction circle piece and one light green piece is the smallest fraction circle piece.

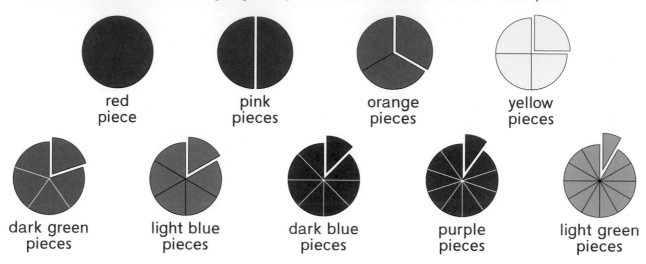

| red piece | pink pieces | orange pieces | yellow pieces |

| dark green pieces | light blue pieces | dark blue pieces | purple pieces | light green pieces |

The fraction name for any of the fraction circle pieces depends on which piece is the whole.

Example

The red fraction circle piece is the whole. Show $\frac{1}{3}$ (one-third) of the red piece.

$\frac{1}{3}$ means one out of three equal parts of the whole. Find three equal-size pieces that cover the whole red piece. Three orange pieces cover the whole red piece, so each orange piece is a third.

One orange piece is $\frac{1}{3}$ (one-third) of the red piece.

Whole
red circle

Example

The yellow fraction circle piece is the whole. Which fraction circle piece is $\frac{1}{3}$ (one-third) of the whole?

Find three equal-size pieces that cover the whole yellow piece. Three light green pieces cover the whole yellow piece, so each light green piece is a third.

One light green piece is $\frac{1}{3}$ (one-third) of the yellow piece.

Whole
yellow piece

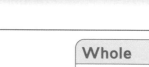

The fraction circle piece that represents $\frac{1}{3}$ is *not* the same when the wholes are different sizes. One-third of the red piece is *greater than* one-third of the yellow piece because the red piece is *larger than* the yellow piece.

Whole
red circle

Representing Numbers Greater Than One

You can use fraction circle pieces to show fractions greater than one. The red circle is the whole for the examples on this page.

Example

Use fraction circle pieces to show $\frac{4}{3}$.

Three orange pieces completely cover a red circle. So each orange fraction circle piece is one-third of the whole, or $\frac{1}{3}$. Four orange pieces show $\frac{1}{3} + \frac{1}{3} + \frac{1}{3} + \frac{1}{3} = 4 * \left(\frac{1}{3}\right)$, or $\frac{4}{3}$.

The four orange pieces can also be used to make as many wholes as possible. Since 3 orange pieces completely cover a red circle, $\frac{3}{3}$ can be combined to show one whole with $\frac{1}{3}$ remaining.

One whole and one-third, or $\frac{4}{3}$, can be named using a mixed number: $1\frac{1}{3}$.

Example

What are two names for the amount shown?

Since 8 equal-size pieces make one whole, each dark blue piece is an eighth of a whole, or $\frac{1}{8}$.

One way:

In all, there are $8 + 8 + 5 = 21$ dark blue pieces.

The amount shown is $\frac{8}{8} + \frac{8}{8} + \frac{5}{8} = \frac{21}{8}$.

Another way:

The 2 complete fraction circles show 2 wholes.

There are 5 more dark blue pieces. That means there is another $\frac{5}{8}$ of a whole.

In all, there are 2 wholes and $\frac{5}{8}$ of a whole. Two and five-eighths can be written as $2\frac{5}{8}$.

$\frac{21}{8}$ and $2\frac{5}{8}$ are two names for the amount shown when the red piece is the whole.

$\frac{21}{8} = 2\frac{5}{8}$

Check Your Understanding

Use fraction circle pieces or drawings to show each of the following. Write another name for the amount shown. The red piece is the whole for each problem.

1. $\frac{8}{6}$ **2.** $1\frac{3}{8}$ **3.** $\frac{10}{4}$ **4.** $2\frac{1}{3}$

Check your answers in the Answer Key.

Using Fraction Strips

A fraction strip is a long rectangle that represents the whole. The fraction strip rectangle on the right is called the *whole*, *one whole*, or *one*.

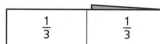

Whole

fraction strip

Fraction strips can be folded into equal-size pieces to represent fractions. A **unit fraction** names one equal part of the whole.

1 whole

This fraction strip is partitioned into thirds. The unit fraction is $\frac{1}{3}$.

$\frac{1}{3}$	$\frac{1}{3}$	$\frac{1}{3}$

You can show amounts less than one by folding the strip.

Example

What fraction is represented by this fraction strip?

The amount shown is greater than 0, but less than one whole fraction strip. Each equal-size piece shows one-third of a whole fraction strip.

$\frac{1}{3}$	$\frac{1}{3}$

Count: 1 third, 2 thirds. Two-thirds of the whole fraction strip is shown.

$$\frac{1}{3} + \frac{1}{3} = \frac{2}{3}$$

You can use more than one fraction strip to show fractions greater than one.

Example

What fraction is represented by these fraction strips?

The amount shown is greater than 1, but less than 2.

$\frac{1}{4}$	$\frac{1}{4}$	$\frac{1}{4}$	$\frac{1}{4}$

$\frac{1}{4}$	$\frac{1}{4}$	$\frac{1}{4}$

Count the unit fractions. There are 7 fourths. The fraction is seven-fourths, or $\frac{7}{4}$.

You can also think about the wholes first, then count on to name the fraction for the remaining parts.

One whole and three-fourths strips are shown. You can write one and three-fourths as $1\frac{3}{4}$.

Using Fractions to Name Points on a Number Line

Number lines are used to show distance. The distance from 0 to 1 on a number line represents one whole. Equal-size distances within a whole are called **equal parts.**

To show a distance traveled on the number line, start at 0, move the given distance, and stop. The stopping point is marked at the end of the distance traveled. You can use fractions to name these points on a number line.

Example

On the number line below, use your finger to trace the distance that starts at 0 and ends at 1. The location of the stopping point, shown by a dot, is 1.

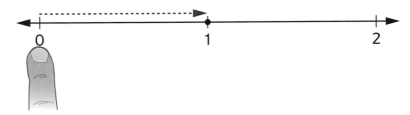

Locating Numbers Between Zero and One

Number lines can be partitioned, or divided into equal parts, to show distances from 0 that are less than 1.

Example

What is the location of point A?

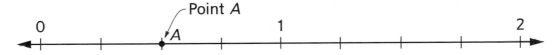

The distance from 0 to point A is less than one whole.

The whole is partitioned into four equal parts, or fourths. Count the fourths starting at the first mark: 0 fourths, 1 fourth, 2 fourths.

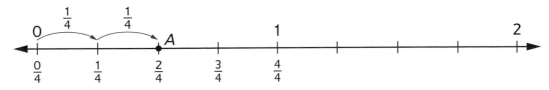

Point A is two-fourths, or $\frac{2}{4}$, the distance from 0 to 1 whole.

Locating Fractions Greater Than One

Number lines can show fractions greater than 1. The distance from 0 to 1 on a number line is one whole. The distance from 0 to 2 is two wholes, the distance from 0 to 3 is three wholes, and so on. Equal-size distances within each whole are equal parts.

Example

What is the location of point *C*?

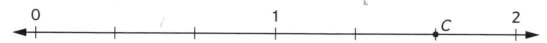

Each whole on the number line is partitioned into 3 equal parts, or thirds. Count the thirds to find the distance of point *C* from 0: 0 thirds, 1 third, 2 thirds, 3 thirds, 4 thirds, 5 thirds.

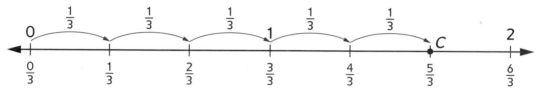

The location of point *C* is $\frac{5}{3}$.

Another way to name the location of point *C* is to first count the number of wholes traveled. Then count on to name the fraction that represents the rest of the distance.

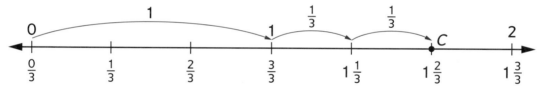

Trace the distance from 0 to 1, and count 1 whole. Then count on by thirds from 1 until you reach point *C*: 1 and 1 third, 1 and 2 thirds. Since point *C* is two-thirds past one whole, the fraction can be named one and two-thirds. It can be written $1\frac{2}{3}$.

$1\frac{2}{3}$ is the same distance from 0 as $\frac{5}{3}$.

Example

What fraction names point *D*?

The number line is partitioned into halves. Find the distance from 0 to point *D* by counting the number of halves: 0 halves, 1 half, 2 halves, 3 halves, 4 halves, 5 halves.

A fraction that names point *D* is $\frac{5}{2}$.

You can also name point *D* with a mixed number. When counting on, point *D* is 2 wholes and another half from 0. Another name for point *D* is two and one-half, or $2\frac{1}{2}$.

Plotting Points on a Number Line

To plot, or locate, a fraction on a number line, draw a number line and label the whole numbers. Use the denominator to partition the whole on the number line into equal-size parts, and then use the numerator to count the number of fractional parts to locate the point.

Example

Plot $\frac{2}{3}$ on a number line.

Start with a number line from 0 to a whole number larger than the fraction.

Since $\frac{2}{3}$ is between 0 and 1, the number line should extend to at least 1.

Look at the denominator. Use tick marks to partition each whole on the number line into equal parts.

The denominator is 3. That means the whole should be partitioned into thirds. Divide the distance between 0 and 1 into 3 equal parts.

Note Make sure to count the equal parts instead of the tick marks between whole numbers. The number of tick marks between whole numbers is one less than the number of parts.

Start at 0. Trace the distance of the fraction you want to locate. At the end of the distance, make a dot for the point and label it.

Trace the distance to $\frac{2}{3}$. Count: 1 third, 2 thirds. At the end of that distance, put a dot for the point at that location and label it $\frac{2}{3}$.

Check Your Understanding

Sketch number lines that show the location of each fraction.

1. $\frac{1}{2}$ **2.** $\frac{3}{4}$ **3.** $\frac{4}{3}$

What fraction names the point on the number line?

4.

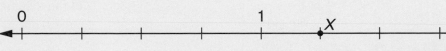

Check your answers in the Answer Key.

Number and Operations—Fractions

Equivalent Fractions

Fractions that name the same amount or the same distance from 0 on a number line are called **equivalent fractions.** Equivalent fractions are equal because they name the same number.

Example

The four circles below are the same size, but they are divided into different numbers of parts. The shaded areas are the same in each circle. These circles show different fractions that are equivalent to $\frac{1}{2}$.

 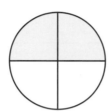  

2 equal parts	4 equal parts	6 equal parts	8 equal parts
1 part shaded	2 parts shaded	3 parts shaded	4 parts shaded
$\frac{1}{2}$ of the circle is shaded.	$\frac{2}{4}$ of the circle is shaded.	$\frac{3}{6}$ of the circle is shaded.	$\frac{4}{8}$ of the circle is shaded.

The fractions $\frac{1}{2}$, $\frac{2}{4}$, $\frac{3}{6}$, and $\frac{4}{8}$ are all equivalent because they represent the same amount of the whole. They are different names for the part of the whole that is shaded.

You can write:

$$\frac{1}{2} = \frac{2}{4} \qquad \frac{1}{2} = \frac{4}{8} \qquad \frac{2}{4} = \frac{4}{8}$$

$$\frac{1}{2} = \frac{3}{6} \qquad \frac{2}{4} = \frac{3}{6} \qquad \frac{3}{6} = \frac{4}{8}$$

Example

On Ms. Cline's bus route, she picks up 12 students. There are 8 boys and 4 girls.

What fraction of the students are girls?

B	B	B	B
B	B	B	B
G	G	G	G

B	B	B	B
B	B	B	B
G	G	G	G

B	B	B	B
B	B	B	B
G	G	G	G

3 equal groups
Each group is $\frac{1}{3}$ of the total.

6 equal groups
Each group is $\frac{1}{6}$ of the total.

12 equal groups
Each group is $\frac{1}{12}$ of the total.

1 group is girls.
$\frac{1}{3}$ of the students are girls.

2 groups are girls.
$\frac{2}{6}$ of the students are girls.

4 groups are girls.
$\frac{4}{12}$ of the students are girls.

The fractions $\frac{1}{3}$, $\frac{2}{6}$, and $\frac{4}{12}$ are equivalent fractions that all name the fraction of the students on the bus that are girls. $\frac{1}{3} = \frac{2}{6} = \frac{4}{12}$

SRB
136 one hundred thirty-six

Finding Equivalent Fractions

You can use fraction tools to find equivalent fractions.

Number lines can show equivalent fractions when the wholes are the same length on each number line. When fractions are equivalent, they are located at the same point on the number line, or are the same distance from 0.

You can use an Equivalent Fractions Poster or the Number-Line Poster on the next page. To find equivalent fractions:

- Make a vertical line through the point for a fraction.

- If the line goes through the point for another fraction, the fractions are equivalent because they are the same distance from 0.

Example

Name a fraction that is equivalent to $\frac{3}{5}$.

The straightedge makes a vertical line that passes through $\frac{3}{5}$ and $\frac{6}{10}$.

$\frac{3}{5}$ and $\frac{6}{10}$ are equivalent fractions because their points on the number line are the same distance from 0.

Write $\frac{3}{5} = \frac{6}{10}$.

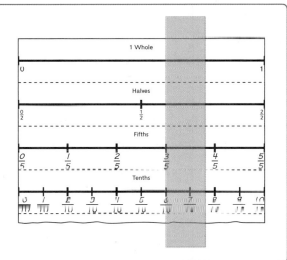

Fraction circles show equivalent fractions when the pieces cover equal areas.

Example

Name a fraction that is equivalent to $\frac{1}{3}$.

If the red circle is the whole, then the orange piece represents $\frac{1}{3}$.

Two light blue pieces cover the same area as the orange piece, so two-sixths is the same amount as one-third.

Two-sixths of this circle covers the same amount even when the sixths are in different places.

$\frac{2}{6}$ is equivalent to $\frac{1}{3}$. Write $\frac{2}{6} = \frac{1}{3}$.

Number-Line Poster

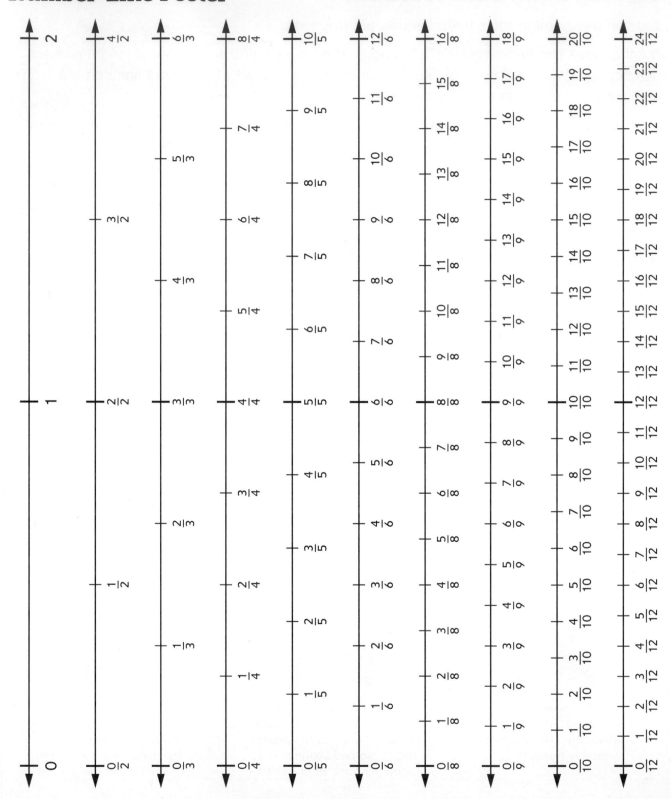

Equivalent Fractions on a Ruler

Inch rulers often have tick marks of different lengths. The longest marks on the ruler below show whole inches. Marks for half inches are shorter than marks for whole inches; marks for quarter inches are shorter than marks for half inches; and so on. The shortest marks show sixteenths of an inch.

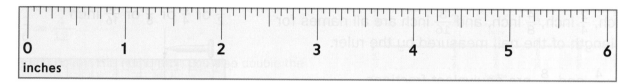

Every tick mark on this ruler can be named by a number of sixteenths. Some tick marks can also be named by eighths, fourths, halves, and wholes. The picture below shows the pattern of fraction names for a part of the ruler.

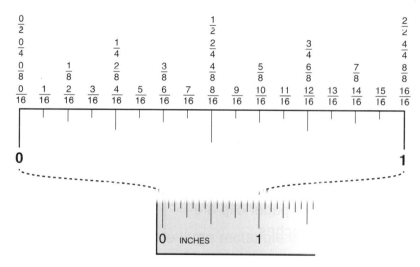

This pattern of naming tick marks continues past 1 inch, but mixed numbers are used to name the length.

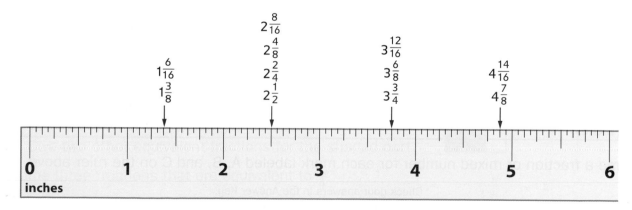

Renaming Mixed Numbers

You can rename **mixed numbers** as fractions greater than one.

Rename $2\frac{3}{5}$ as a fraction.

One way: Think about breaking apart the wholes.

If a red circle is the whole, then $2\frac{3}{5}$ is two whole circles and $\frac{3}{5}$ of a third circle.

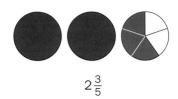

$2\frac{3}{5}$

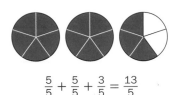

$\frac{5}{5} + \frac{5}{5} + \frac{3}{5} = \frac{13}{5}$

If you break apart each of the whole circles into fifths, then you can see that $2\frac{3}{5} = \frac{13}{5}$.

Another way: Think about a number line.

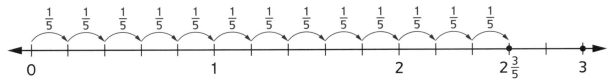

Count the fifths. You count 13 fifths from 0 to $2\frac{3}{5}$, so $2\frac{3}{5} = \frac{13}{5}$.

Another way: Rename the whole number as a fraction with the same denominator as the fraction part, and add the fractions.

$2\frac{3}{5} = 2 + \frac{3}{5}$

Rename 1 as $\frac{5}{5}$.

Since $1 = \frac{5}{5}$, $2 = \frac{5}{5} + \frac{5}{5}$, or $\frac{10}{5}$.

So, $2\frac{3}{5} = \frac{10}{5} + \frac{3}{5} = \frac{13}{5}$.

Note You can think of making wholes and breaking apart wholes as fair trades. Putting 4 fourths together to make a whole is like trading $\frac{4}{4}$ for 1. This is a fair trade because $\frac{4}{4} = 1$. Breaking apart a whole into 5 fifths is like trading 1 for $\frac{5}{5}$.

Write each fraction as a mixed number. Write each mixed number as a fraction.

1. $\frac{11}{4}$ 2. $\frac{20}{8}$ 3. $3\frac{5}{8}$ 4. $2\frac{20}{100}$

Check your answers in the Answer Key.

Comparing Fractions

You can compare fractions when they name parts of the same whole. You can use fraction tools to compare fractions.

Example

Compare $\frac{5}{8}$ and $\frac{3}{4}$.

Find $\frac{5}{8}$ and $\frac{3}{4}$ on the Number-Line Poster.

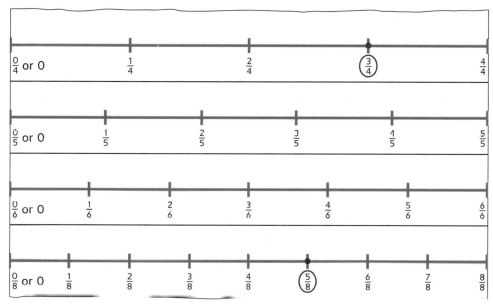

The whole is the same distance from 0 on both number lines.

$\frac{3}{4}$ is to the right of $\frac{5}{8}$, so $\frac{3}{4}$ is a greater distance from 0 than $\frac{5}{8}$.

Write $\frac{3}{4} > \frac{5}{8}$ or $\frac{5}{8} < \frac{3}{4}$.

Example

Which is greater, $\frac{2}{3}$ or $\frac{7}{8}$?

You can use the red fraction circle as the whole.

Show $\frac{2}{3}$ and $\frac{7}{8}$ with fraction circle pieces.

<	is less than
>	is greater than
=	is equal to

$\frac{2}{3}$

$\frac{7}{8}$

Two-thirds covers less space than seven-eighths. $\frac{2}{3}$ is *less than* $\frac{7}{8}$, so $\frac{7}{8}$ is *greater than* $\frac{2}{3}$. Write $\frac{2}{3} < \frac{7}{8}$ or $\frac{7}{8} > \frac{2}{3}$.

Number and Operations—Fractions

Example

Kiara and Tania each had a small pizza for lunch. The pizzas were the same size when they started eating. Kiara ate $\frac{5}{6}$ of her pizza. Tania ate $\frac{3}{4}$ of her pizza. Who ate more pizza?

You can model the part of the pizza that each girl ate using fraction circle pieces. A red fraction circle can be used to represent the whole pizza. The fraction circle pieces show the amount of pizza each girl ate. The empty parts of the circles show the amount that each girl *did not* eat. The piece of pizza that Tania *did not eat* is larger than the piece of pizza that Kiara *did not eat*. That means the amount that Kiara ate is *closer to a whole pizza*.

Kiara   Tania

Kiara ate more pizza.

The size of a fraction depends on both the numerator and denominator, so when you compare fractions, you have to pay attention to both the numerator and the denominator.

Note Fractions with **like denominators** have the same denominator.

$\frac{1}{4}$ and $\frac{3}{4}$ have like denominators.

Fractions with **like numerators** have the same numerator.

$\frac{2}{3}$ and $\frac{2}{5}$ have like numerators.

Comparing Fractions with Like Denominators

When two fractions of the same-size whole have the same **denominator,** it means that the parts that the whole is divided into are the same size for both fractions. So you can compare numerators, or the number of parts.

Example

Decide whether $\frac{7}{8}$ or $\frac{5}{8}$ is larger.

Because the eighths are the same size, 7 eighths is more than 5 eighths. You can think of plotting $\frac{7}{8}$ and $\frac{5}{8}$ on a number line where the whole is partitioned into eighths.

$\frac{7}{8}$ is to the right of $\frac{5}{8}$ because it is a longer distance from 0. So the fraction with the larger numerator is the larger number. Because $\frac{5}{8}$ is a shorter distance from 0 than $\frac{7}{8}$, the fraction with the smaller numerator is the smaller number.

$\frac{7}{8}$ is larger than $\frac{5}{8}$. Write $\frac{7}{8} > \frac{5}{8}$ or $\frac{5}{8} < \frac{7}{8}$.

Examples

Compare $\frac{4}{5}$ and $\frac{3}{5}$.

$\frac{4}{5} > \frac{3}{5}$ because $4 > 3$ and the denominators are the same.

Compare $\frac{2}{9}$ and $\frac{7}{9}$.

$\frac{2}{9} < \frac{7}{9}$ because $2 < 7$ and the denominators are the same.

Comparing Fractions with Like Numerators

When fractions of the same-size whole have the same **numerator,** you can think about the size of the denominators to decide which fraction is larger. A smaller denominator means the whole is divided into fewer equal parts, so each part is bigger. A larger denominator means the whole is divided into more equal parts, so each part is smaller. When the numerators of the fractions are the same, the fraction with the smaller denominator is the greater fraction.

Example

Compare $\frac{3}{5}$ and $\frac{3}{8}$.

Fifths are bigger than eighths when the whole is the same size.

So, 3 fifths is greater than 3 eighths.

$\frac{3}{5}$ $\frac{3}{8}$

$$\frac{3}{5} > \frac{3}{8} \text{ or } \frac{3}{8} < \frac{3}{5}$$

Examples

Compare $\frac{1}{2}$ and $\frac{1}{3}$.

$\frac{1}{2} > \frac{1}{3}$ because halves are bigger than thirds when the whole is the same.

Compare $\frac{3}{8}$ and $\frac{3}{4}$.

$\frac{3}{8} < \frac{3}{4}$ because eighths are smaller than fourths when the whole is the same.

Check Your Understanding

Use fraction tools, drawings, or reasoning to compare. Use <, >, or =.

1. $\frac{3}{5} \square \frac{3}{10}$ **2.** $\frac{2}{3} \square \frac{3}{8}$ **3.** $\frac{3}{8} \square \frac{5}{8}$ **4.** $\frac{2}{6} \square \frac{2}{5}$

Check your answers in the Answer Key.

Number and Operations—Fractions

Comparing Fractions with Unlike Numerators and Unlike Denominators

You can choose from several strategies to compare fractions when both the numerators and the denominators are different.

Comparing to Benchmarks. A **benchmark** is a familiar reference point. Numbers such as 0, $\frac{1}{2}$, 1, $1\frac{1}{2}$, and 2 are often used as benchmarks because they are easy to visualize, or picture in your head. You can often compare two fractions by first comparing each fraction to a common benchmark.

Example

Compare $\frac{2}{5}$ and $\frac{5}{8}$.

Think: How does each fraction compare to $\frac{1}{2}$?

Notice that $\frac{5}{8}$ is more than $\frac{1}{2}$ and $\frac{2}{5}$ is less than $\frac{1}{2}$.

So, $\frac{2}{5} < \frac{5}{8}$ and $\frac{5}{8} > \frac{2}{5}$.

Example

Compare $\frac{7}{8}$ and $\frac{3}{4}$.

Think: Both fractions are less than 1. Which fraction is closer to 1?

$\frac{7}{8}$ is $\frac{1}{8}$ away from 1. $\frac{3}{4}$ is $\frac{1}{4}$ away from 1. Since eighths are smaller than fourths, $\frac{7}{8}$ is closer to 1.

Write $\frac{7}{8} > \frac{3}{4}$ or $\frac{3}{4} < \frac{7}{8}$.

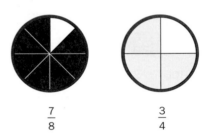

$\frac{7}{8}$ $\frac{3}{4}$

Using Equivalent Fractions. Sometimes you need to compare fractions with unlike denominators that may be difficult to compare using benchmarks or representations. You can always find equivalent fractions that have the same denominator and then compare.

Example

Compare $\frac{2}{3}$ and $\frac{5}{6}$.

One way:

Look at the table of equivalent fractions on page 142. Find fractions with like denominators for $\frac{2}{3}$ and $\frac{5}{6}$.

The table shows that $\frac{2}{3} = \frac{8}{12}$ and $\frac{5}{6} = \frac{10}{12}$.

Compare $\frac{8}{12}$ and $\frac{10}{12}$.

Since $\frac{8}{12} < \frac{10}{12}$, you know that $\frac{2}{3} < \frac{5}{6}$.

Another way:

Rename $\frac{2}{3}$ as sixths using the multiplication rule.

$$\frac{2}{3} = \frac{(2 * 2)}{(2 * 3)} = \frac{4}{6}$$

Compare $\frac{4}{6}$ and $\frac{5}{6}$.

Since $\frac{4}{6} < \frac{5}{6}$, you know that $\frac{2}{3} < \frac{5}{6}$.

Introducing Decimals

Decimals, like fractions, are used to name numbers that are between whole numbers. Every fraction can be renamed as a decimal. While fractions are written using fraction notation, decimals are written using the **base-10 place-value** system. That means that you can compare and compute decimals in similar ways as you do whole numbers.

You probably see many uses of decimals every day.

Metric Measure	Decimals can be used to name lengths measured with metric rulers, metersticks, and tape measures.	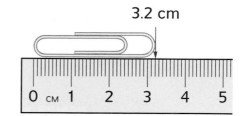
	The centimeter ruler shows that the paper clip is 3.2 (3 wholes and 2 tenths) centimeters long.	

Money

Fractional parts of a dollar are almost always written as decimals.

The receipt shows that lunch cost between 25 dollars and 26 dollars. The "64" in the total cost names a part of a dollar: $\frac{64}{100}$ of a dollar, or 64 cents.

```
   THE  QUAY   KEY  WEST
^^^^^^^^^^^^^^^^^^^^^^^

0127     Table 26  #Party 2
TOM K       SvrCk: 7 15:14 07/11/05

1 ICED TEA                    1.25
1 CALIF BURGER                7.25
1 CHEFS SANDWICH              7.95
1 COFFEE                      1.50
2 KEY LIME PIE                5.90

                 Sub Total:  23.85
                       Tax    1.77
07/11 15:51   TOTAL:   25.84

********************

WE HOPE YOU ENJOY YOUR STAY IN
          KEY WEST.


      COME  BACK  SOON!
```

Body Temperature and Other Measures

Decimals are used to report body temperature and other measures such as weight.

Normal body temperature is around 98.6°F.

Understanding Decimals

Decimals are another way to write fractions. Many fractions have denominators of 10, 100, 1,000, and so on. In a decimal, the whole number is written to the left of a dot called a **decimal point.** Digits to the right of the decimal point show parts of a whole.

decimal point

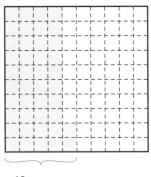

0.27

whole part of
number a whole

In the pictures below, the square is the whole. The fraction and decimal name for the amount shaded in each square is written below each one.

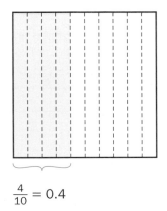

$\frac{4}{10} = 0.4$

This square is divided into 10 equal parts. Each part is $\frac{1}{10}$ of the square. When one digit is written to the right of the decimal point, it tells you that the name of the parts is tenths. So the decimal name for $\frac{1}{10}$ is 0.1.

Since $\frac{4}{10}$ of the square above is shaded, the decimal name is:

$$\underbrace{0}_{\text{wholes}} \quad . \quad \underbrace{4}_{\text{tenths}}$$

$\frac{42}{100} = 0.42$

This square is divided into 100 equal parts. Each part is $\frac{1}{100}$ of the square. When two digits are written to the right of the decimal point, it tells you that the name of the parts is hundredths. So the decimal name for $\frac{1}{100}$ is 0.01.

Since $\frac{42}{100}$ of the square above is shaded, the decimal name is:

$$\underbrace{0}_{\text{wholes}} \quad . \quad \underbrace{42}_{\text{hundredths}}$$

Like mixed numbers, decimals can name numbers greater than 1.

Example

What is the amount shown?

$2\frac{34}{100} = 2.34$

The amount shown is two and thirty-four hundredths.

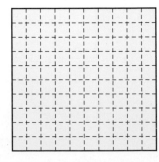 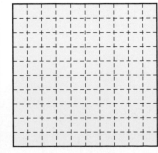 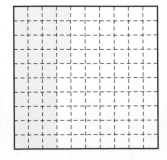

Renaming a Fraction as a Decimal

You can rename a fraction as a decimal if you can find an equivalent fraction with a denominator that is a multiple of 10, such as 10 or 100. For example, $\frac{3}{10} = 0.3$ and $\frac{72}{100} = 0.72$.

Renaming a Decimal as a Fraction

To change a decimal to a fraction, write the decimal as a fraction with a denominator of 10 or 100.

Examples

Write each decimal as a fraction.

For 0.5, the rightmost digit is 5, which is in the tenths place. So, 0.5 is the same as 5 tenths, or $\frac{5}{10}$.

For 4.92, the rightmost digit is 2, which is in the hundredths place. So, $4.92 = \frac{442}{100}$, or $4\frac{92}{100}$. You can say 4 wholes and 92 hundredths, or 4 and 92 hundredths.

Reading Decimals

The **decimal point** separates the whole number and the part of a whole. This helps you read the number. Use these guidelines to read the decimal aloud.

- Non-zero digits to the left of the decimal point are read just like a whole number.

- Follow the whole number name with the word *and* to locate the decimal point.

- Then read the digits to the right like a fraction, where the number to the right tells how many parts, and you say the size of the part by saying:

  - *tenths* if there is exactly one digit to the right of the decimal point

  - *hundredths* if there are exactly two digits to the right of the decimal point

Examples

Read the decimals.

104.6

The 104 to the left of the **decimal point** is read as "one hundred four." Use the word *and* to give the location of the decimal point. The 6 is one digit to the right of the decimal point. It is read as "six-*tenths*."

104.6 is read as "one hundred four and six-tenths."

0.03

The zero to the left of the **decimal point** means there are no wholes, so don't say anything for the whole number. The 03 is two digits to the right of the decimal point, so it is read "three-*hundredths*."

0.03 is read as "three-hundredths."

Extending Place Value to Decimals

The base-10 system works the same way for decimals as it does for whole numbers.

Examples

1,000s thousands	100s hundreds	10s tens	1s ones	.	0.1s tenths	0.01s hundredths
		3	6	.	7	4
			3	.	0	5

In the number 36.74,

7 is in the **tenths** place; its value is 7 tenths, or $\frac{7}{10}$, or 0.7.

4 is in the **hundredths** place; its value is 4 hundredths, or $\frac{4}{100}$, or 0.04.

In the number 3.05,

0 is in the **tenths** place; its value is 0.

5 is in the **hundredths** place; its value is 5 hundredths, or $\frac{5}{100}$, or 0.05.

Study the place-value chart below. Look at the numbers that name the places. A digit in one place represents **10 times as much as it represents in the place to its right**.

$$*10 \quad *10 \quad *10 \quad *10 \quad *10$$

1,000s thousands	100s hundreds	10s tens	1s ones	.	0.1s tenths	0.01s hundredths

$$1\left[\frac{1}{10}\right] = 10\left[\frac{1}{100}s\right]$$
$$1[1] = 10\left[\frac{1}{10}s\right]$$
$$1[10] = 10[1s]$$
$$1[100] = 10[10s]$$
$$1[1,000] = 10[100s]$$

Check Your Understanding

For each picture, write a decimal and write the words you would use to read the decimal.

1.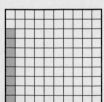

2.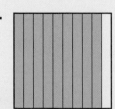

Write the words you would use to read each decimal.

3. 24.68   **4.** 4.06

Check your answers in the Answer Key.

You can use facts about the base-10 place-value chart to help you make trades using base-10 blocks.

Example

Suppose that a flat is worth 1. Then a long is worth $\frac{1}{10}$, or 0.1, and a cube is worth $\frac{1}{100}$, or 0.01.

You can trade one flat for 100 cubes because one whole is equal to one hundred $\frac{1}{100}$s.

You can trade one flat for ten longs because one whole is equal to ten $\frac{1}{10}$s.

You can trade one long for ten cubes because one $\frac{1}{10}$ is equal to ten $\frac{1}{100}$s.

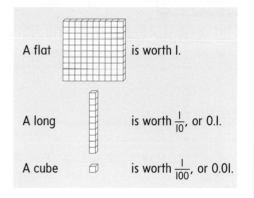

A flat   is worth 1.

A long   is worth $\frac{1}{10}$, or 0.1.

A cube   is worth $\frac{1}{100}$, or 0.01.

Study the place value chart below. Look at the numbers that name the places. A digit in one place represents $\frac{1}{10}$ **of what it represents in the place to its left.** Finding $\frac{1}{10}$ of a number is the same as dividing that number by 10.

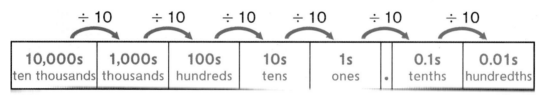

10,000s	1,000s	100s	10s	1s		0.1s	0.01s
ten thousands	thousands	hundreds	tens	ones	.	tenths	hundredths

---

Check Your Understanding

1. What is the value of the digit 2 in each of these numbers?

   a. 204.7   **b.** 0.21   **c.** 4.12   **d.** 512.11

2. Tell how much each digit in 87.65 is worth.

Check your answers in the Answer Key.

Did You Know?

People in the United States, Australia, and most Asian countries write the decimal 3.25. In the United Kingdom, the decimal is written 3·25 and in some parts of Europe and South America, the decimal is written 3,25.

Comparing Decimals

One way to compare decimals is to model them with base-10 blocks.

Example

Compare 0.27 and 0.3. The flat is the whole.

0.27 is 2 tenths and 7 hundredths.   0.3 is 3 tenths.

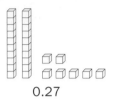
0.27

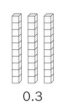
0.3

2 longs and 7 cubes are less than 3 longs.

So, 0.27 is less than 0.3. 0.27 < 0.3

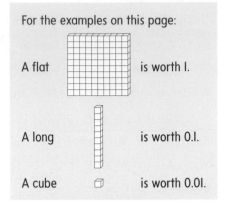

For the examples on this page:

A flat   is worth I.

A long   is worth 0.I.

A cube   is worth 0.0I.

Another way to compare decimals is to write them as fractions with a like denominator.

Example

Compare 0.6 and 0.71.

Write 0.6, or six-tenths, as a fraction: $\frac{6}{10}$. Write 0.71, or seventy-one hundredths, as a fraction: $\frac{71}{100}$.

Write $\frac{6}{10}$ as a fraction with a denominator of 100: $\frac{6}{10} = \frac{(6 * 10)}{(10 * 10)} = \frac{60}{100}$

$\frac{71}{100} > \frac{60}{100}$ Seventy-one hundredths is more than sixty-hundredths.

0.71 > 0.60, so 0.6 < 0.71.

You can trade a long for 10 cubes because $\frac{1}{10} = \frac{10}{100}$. You can write 0.1 = 0.10 in decimal notation. You can write a 0 at the end of a decimal without changing the value of the decimal: 0.7 = 0.70 and 0.4 = 0.40. Think of it as trading for equivalent amounts that use smaller pieces.

Example

Show that 0.3 = 0.30.

One way:
Use base-10 blocks. A flat is the whole.

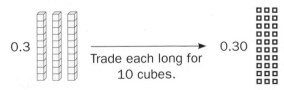

0.3   Trade each long for 10 cubes.   0.30

Another way:
Write 0.3, or 3 tenths, as a fraction: $\frac{3}{10}$.
Write 0.30, or 30 hundredths, as a fraction: $\frac{30}{100}$.

Write $\frac{3}{10}$ as an equivalent fraction with a denominator of 100:

$\frac{3}{10} = \frac{(3 * 10)}{(10 * 10)} = \frac{30}{100}$

$\frac{3}{10}$ is equivalent to $\frac{30}{100}$.

0.3 = 0.30

Writing a 0 at the end of a decimal without changing the value of the decimal can make comparing decimals easier.

Example

Compare 0.2 and 0.05. A flat is the whole.

0.2 is 2 tenths.   0.05 is 5 hundredths.

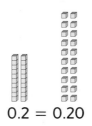

0.2 = 0.20   0.05

Trade 2 longs for 20 cubes.

20 cubes is more than 5 cubes. 20 hundredths is more than 5 hundredths.

0.20 > 0.05, so 0.2 > 0.05.

Example

Compare 0.99 and 1. A flat is the whole.

Think about base-10 blocks. 1 = 1.00

1.00 is 1 flat, or 100 cubes, or $\frac{100}{100}$. 0.99 is 99 cubes, or $\frac{99}{100}$.

$\frac{99}{100}$ is less than $\frac{100}{100}$, so 0.99 < 1.00, or 1.

A place-value chart can also help you compare decimals.

Example

Compare 13.82 and 13.76.

The tens digits *are the same*. They are both worth 10.
The ones digits *are the same*. They are both worth 3.
The tenths digits *are not the same*.
The 8 is worth 8 tenths, or 0.8.
The 7 is worth 7 tenths, or 0.7.
8 tenths is worth more than 7 tenths.

10s tens	1s ones	•	0.1s tenths	0.01s hundredths
1	3	.	8	2
1	3	.	7	6

So, 13.82 is worth more than 13.76. 13.82 > 13.76

Check Your Understanding

Insert >, <, or = to make a true number sentence.

1. 0.68 _____ 0.2   **2.** 5.39 _____ 5.5   **3.** $\frac{1}{2}$ _____ 0.51

4. 0.60 _____ 0.6   **5.** 0.44 _____ 0.4

Check your answers in the Answer Key.

Problem Solving by Drawing and Reasoning about Fractions

You can use pictures, fraction circles, fraction strips, fraction number lines, and mathematical reasoning to help you solve fraction problems.

Examples

Describe the whole for each situation.

Derek ran $\frac{3}{4}$ of the way home from school.

The whole is the entire distance from school to home. The fraction, $\frac{3}{4}$, names the part of the distance that Derek ran.

In Mrs. Blake's classroom, $\frac{1}{2}$ of the students are girls.

The whole is the collection of all students in Mrs. Blake's classroom. The fraction, $\frac{1}{2}$, names the part of that collection that are girls.

Example

Sally found half a leftover pizza in the refrigerator. Is that a lot?

The answer depends on how big the pizza is.

Think about pizzas of different sizes.

Draw two different-size pizzas.

One-half of a large pizza is a more than one-half of a small pizza.

Example

Four sisters share this collection of 12 buttons fairly. What fraction of the buttons does each sister get?

The whole is the collection of buttons.

Each button is a part of the collection, and there are 12 buttons in all. So the denominator is 12.

Since 4 [3s] = 12, each sister will get 3 buttons. So the numerator is 3.

Each sister gets $\frac{3}{12}$ of the collection of buttons.

Whole
12 buttons

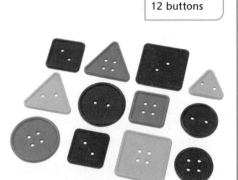

Example

A recipe requires $1\frac{1}{4}$ cups of flour. You only have a $\frac{1}{4}$-cup measuring cup. How can you put the correct amount of flour in the recipe?

Imagine using fraction circle pieces to decide how many $\frac{1}{4}$s make $1\frac{1}{4}$. Think of the red circle as the whole.

Since $\frac{4}{4}$ makes one full circle and you need $1\frac{1}{4}$ in all, you need another 1 fourth, or 5 fourths in all.

Measure $\frac{1}{4}$ cup of flour 5 times to get $1\frac{1}{4}$ cups.

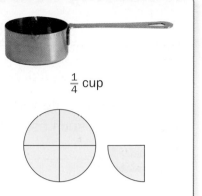

$\frac{1}{4}$ cup

Example

Two people want to share 3 apples equally. How much apple will each person get?

One way: Draw 3 apples cut into 2 equal pieces, or 6 halves. Then show giving 1 half of each apple to each person.

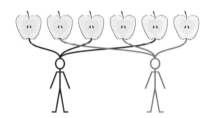

Each person gets 3 pieces of apple. That is 3 halves of an apple per person.

Another way: First draw 2 whole apples and show giving 1 apple to each person. Then show the last apple cut into 2 equal parts.

Each person gets a whole apple and then an additional one-half of an apple. Each person gets one and one-half apples.

Either way, each person gets the same amount of apple. Three-halves of an apple is the same as $1\frac{1}{2}$ apples. $\frac{3}{2} = 1\frac{1}{2}$

Check Your Understanding

Use mathematical reasoning, pictures, or fraction tools to solve these problems.

1. Jenny and Ellen each saved $\frac{1}{2}$ of the money they got for their birthdays. Did they save same amount? Why or why not?

2. Ten dogs are given equal shares of 2 pounds of food. How many pounds of food does each dog get?

3. Art, Jerry, and Lenny shared 2 same-size pizzas equally. How much of a pizza did each get?

Check your answers in the Answer Key.

University of Chicago

Estimating with Fractions

When you work with fractions, you can use **estimates** to help make sense of a situation, approximate a calculation, and check that answers are reasonable.

Using Visual Representations

One way to estimate with fractions is to think about visual representations. You can picture fraction tools, such as fraction circles, fraction strips, and number lines, in your head.

Example

Estimate: Will the sum of $\frac{4}{5} + \frac{1}{10}$ be greater than or less than 1?

You can picture each amount with fraction circles.
Use the red circle as the whole.

Picture $\frac{4}{5}$ as 4 dark green pieces. $\frac{4}{5}$ is almost 1 whole.

$\frac{4}{5}$

Picture $\frac{1}{10}$ as 1 purple piece. $\frac{1}{10}$ is a small sliver.

$\frac{1}{10}$

Adding the small $\frac{1}{10}$ sliver to $\frac{4}{5}$ does not complete the whole circle, so the sum will be less than 1.

$\frac{4}{5} + \frac{1}{10}$

So, $\frac{4}{5} + \frac{1}{10} < 1$.

Example

Estimate: Will the sum of $\frac{7}{10} + \frac{3}{4}$ be greater than or less than 1?

You can visualize each fraction on a number line.

$\frac{7}{10}$ is between $\frac{1}{2}$ and 1.

$\frac{3}{4}$ is between $\frac{1}{2}$ and 1.

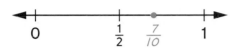

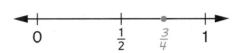

Adding a number greater than $\frac{1}{2}$ to another number greater than $\frac{1}{2}$ will equal a sum greater than 1.

So, the sum of $\frac{7}{10}$ and $\frac{3}{4}$ will be greater than 1.

Example

Estimate: Will the result of $2\frac{2}{6} - \frac{5}{6}$ be greater than or less than 1?

Visualize the starting amount, $2\frac{2}{6}$, using a long rectangle as one whole.

Think about the amount being subtracted. $\frac{5}{6}$ is a little less than 1 whole, so about 1 whole rectangle will be subtracted. Think about crossing out about $\frac{5}{6}$. That will leave at least 1 other whole rectangle and the $\frac{2}{6}$ rectangle remaining.

So, $2\frac{2}{6} - \frac{5}{6}$ will be greater than 1.

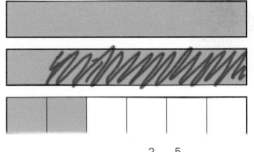

Picturing $2\frac{2}{6} - \frac{5}{6}$

Check Your Understanding

1. Is the sum of $\frac{3}{4} + \frac{1}{2}$ greater than or less than 1? Explain how you used estimation.

2. Is the result of $1\frac{5}{8} - \frac{9}{10}$ greater than or less than 1? Explain how you used estimation.

Check your answers in the Answer Key.

Adding and Subtracting Fractions with Like Denominators

When adding or subtracting fractions, it often helps to use visual representations such as fraction circles, a number line, or a picture. Ask yourself: *Are all of the parts the same size?* If all of the parts are the same size, or have **like denominators,** then you can add or subtract the number of parts given by the numerators.

Example

What is the sum of $\frac{4}{12} + \frac{3}{12}$?

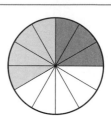

Think about the fractions in terms of the size of their equal parts.
Do both fractions have the same denominator? Yes, both fractions divide the whole into 12 equal-size parts, or twelfths.
Think about the fractions as the sum of unit fractions.

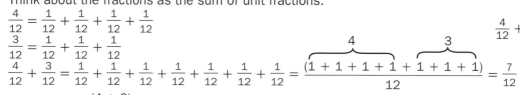

$$\frac{4}{12} = \frac{1}{12} + \frac{1}{12} + \frac{1}{12} + \frac{1}{12}$$

$$\frac{3}{12} = \frac{1}{12} + \frac{1}{12} + \frac{1}{12}$$

$$\frac{4}{12} + \frac{3}{12} = \frac{1}{12} + \frac{1}{12} + \frac{1}{12} + \frac{1}{12} + \frac{1}{12} + \frac{1}{12} + \frac{1}{12} = \frac{\overbrace{(1 + 1 + 1 + 1}^{4} + \overbrace{1 + 1 + 1)}^{3}}{12} = \frac{7}{12}$$

$$\frac{4}{12} + \frac{3}{12} = \frac{(4 + 3)}{12} = \frac{7}{12}$$

You can write $\frac{(4 + 3)}{12}$, or $\frac{7}{12}$.

Example

$\frac{7}{8} + \frac{3}{8} = ?$

Estimate: $\frac{7}{8}$ is close to 1, and $\frac{3}{8}$ is close to $\frac{1}{2}$. $1 + \frac{1}{2} = 1\frac{1}{2}$, so the sum should be close to $1\frac{1}{2}$.

You can represent the problem using rectangles. A long rectangle is one whole.

Think: Are all of the parts the same size?

Yes, both rectangles are divided into eighths.

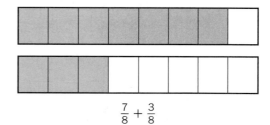

Add the shaded parts of the two rectangles:
7 eighths plus 3 eighths equals 10 eighths.

$$\frac{7}{8} + \frac{3}{8} = \frac{\mathbf{10}}{\mathbf{8}}, \text{ or } \mathbf{1\frac{2}{8}}$$

$$\frac{7}{8} + \frac{3}{8}$$

This makes sense because $1\frac{2}{8}$ is reasonably close to the estimate of $1\frac{1}{2}$.

You can also solve many subtraction problems with fractions by thinking about representations.

Example

Solve $\frac{5}{6} - \frac{2}{6}$.

Estimate: $\frac{5}{6}$ is close to 1. $\frac{2}{6}$ is close to $\frac{1}{2}$. Subtracting a number close to $\frac{1}{2}$ from a number close to 1 will result in an answer close to $\frac{1}{2}$.

One way:
Represent the problem using rectangles.
Start by shading $\frac{5}{6}$ of a partitioned rectangle.

$\frac{2}{6}$ is the same as $\frac{1}{6} + \frac{1}{6}$. Take away 2 parts of $\frac{1}{6}$ each to subtract $\frac{2}{6}$.

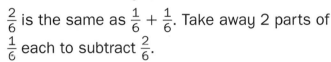

Since both the starting amount and the amount being subtracted are sixths, you can think about subtracting same-size pieces:
5 sixths − 2 sixths = 3 sixths.
Name the result: 3 sixths = $\frac{3}{6}$.

Another way:
Represent the problem on a number line.
Start at $\frac{5}{6}$.

$\frac{2}{6}$ is the same as $\frac{1}{6} + \frac{1}{6}$. Subtracting $\frac{1}{6}$ and another $\frac{1}{6}$ is the same as subtracting $\frac{2}{6}$.

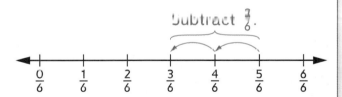

Think: 5 sixths − 2 sixths = 3 sixths.
Name the result: 3 sixths = $\frac{3}{6}$.

So, $\frac{5}{6} - \frac{2}{6} = \frac{3}{6}$. This makes sense because $\frac{3}{6}$ is the same as $\frac{1}{2}$, which was the estimate.

Check Your Understanding

Solve. Explain how you know your answer for Problem 2 makes sense.

1. $\frac{1}{4} + \frac{1}{4} + \frac{1}{4} = ?$ **2.** $\frac{6}{8} - \frac{2}{8} = ?$ **3.** $\frac{3}{4} + \frac{1}{4} = ?$ **4.** $\frac{11}{10} - \frac{2}{10}$

Check your answers in the Answer Key.

Adding Mixed Numbers with Like Denominators

One way to add **mixed numbers** with like denominators is to add the fractions and the whole numbers separately. Make sure you are working with same-size pieces before you add the fractions. You can rename the sum as an **equivalent** mixed number to make your answer easier to check.

Example

$1\frac{5}{8} + 2\frac{1}{8} = $ **?**

Estimate: $1\frac{5}{8}$ is close to $1\frac{1}{2}$. $2\frac{1}{8}$ is close to 2. $1\frac{1}{2} + 2 = 3\frac{1}{2}$, so the sum should be close to $3\frac{1}{2}$.

Add the whole numbers.

$$\begin{array}{r} \mathbf{1}\frac{5}{8} \\ + \mathbf{2}\frac{1}{8} \\ \hline \mathbf{3} \end{array}$$

Add the fractions.

$$\begin{array}{r} 1\frac{5}{8} \\ + 2\frac{1}{8} \\ \hline 3\frac{6}{8} \end{array}$$

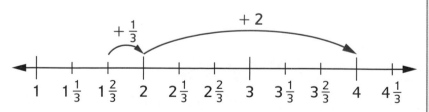

$1\frac{5}{8} + 2\frac{1}{8} = 3 + \frac{(5+1)}{8} = 3\frac{6}{8}$

$1\frac{5}{8} + 2\frac{1}{8} = \mathbf{3\frac{6}{8}}$

This makes sense because $3\frac{6}{8}$ is close to the estimate of $3\frac{1}{2}$.

You can also add mixed numbers by counting up.

Example

$1\frac{2}{3} + 2\frac{1}{3} = $ **?**

Estimate: $1\frac{2}{3}$ is a little less than 2. $2\frac{1}{3}$ is a little more than 2. The sum should be about 4.

Represent $1\frac{2}{3}$ on a number line.
Count up $2\frac{1}{3}$. You land at 4.
$1\frac{2}{3} + 2\frac{1}{3} = \mathbf{4}$
This matches the estimate.

When adding mixed numbers, you will get the same answer whether you add the whole number or fraction parts first.

Example

$2\frac{2}{6} + 1\frac{5}{6} = ?$

Estimate: $2\frac{2}{6}$ is a little less than $2\frac{1}{2}$, and $1\frac{5}{6}$ is almost 2. The sum should be less than $2\frac{1}{2} + 2$, or $4\frac{1}{2}$.

One way:

Add the fractions.

$$2\frac{2}{6}$$
$$+ 1\frac{5}{6}$$
$$\overline{\frac{7}{6}}$$

Add the whole numbers.

$$\mathbf{2}\frac{2}{6}$$
$$+ \mathbf{1}\frac{5}{6}$$
$$\overline{\mathbf{3}\frac{7}{6}}$$

$$2$$
$$+ 1 \qquad\qquad\qquad\qquad + \frac{5}{6}$$
$$\overline{3 \qquad\qquad\qquad\qquad\qquad \frac{7}{6}}$$

Write the mixed number so the fraction part is less than 1: $\frac{7}{6} = \frac{6}{6} + \frac{1}{6}$, or $1\frac{1}{6}$.

So, $3\frac{7}{6}$ is $3 + 1\frac{1}{6}$, or $\mathbf{4\frac{1}{6}}$.

Another way:

Decompose the mixed numbers into the sum of a whole number and a fraction.

Rewrite the whole numbers as fractions.

$$2\frac{2}{6} = 2 + \frac{2}{6} \qquad 1\frac{5}{6} = 1 + \frac{5}{6}$$
$$2 = \frac{12}{6} \qquad\qquad 1 = \frac{6}{6}$$

$$\left(\frac{12}{6} + \frac{2}{6}\right) + \left(\frac{6}{6} + \frac{5}{6}\right)$$

Rewrite the problem using fractions.

$$\frac{14}{6} \qquad + \qquad \frac{11}{6} \qquad = \frac{25}{6} \text{ or } 4\frac{1}{6}$$

$$2\frac{2}{6} + 1\frac{5}{6} = \frac{25}{6} = \mathbf{4\frac{1}{6}}$$

$4\frac{1}{6}$ is less than $4\frac{1}{2}$, which matches the estimate.

Check Your Understanding

Estimate and add.

1. $4\frac{1}{5} + 2\frac{3}{5} = ?$ **2.** $2\frac{6}{8} + 1\frac{4}{8} = ?$

Check your answers in the Answer Key.

Subtracting Mixed Numbers with Like Denominators

When subtracting **mixed numbers,** you can subtract the fractions and whole numbers separately. Check to see whether the fractions have the same denominator before you subtract them.

Example

$4\frac{3}{4} - 2\frac{1}{4} = $ **?**

Estimate: $4\frac{3}{4}$ is close to 5. $2\frac{1}{4}$ is close to 2. $5 - 2 = 3$, so the difference is about 3.

Decompose the mixed numbers into wholes and fractions.

$$4\frac{3}{4} = 4 + \frac{3}{4}$$
$$2\frac{1}{4} = 2 + \frac{1}{4}$$

Subtract the fractions.

$$\begin{array}{r} 4\frac{3}{4} \\ -\ 2\frac{1}{4} \\ \hline \frac{2}{4} \end{array}$$

Subtract the whole numbers.

2 *wholes* and 2 *fourths* remain.

$$\begin{array}{r} \mathbf{4\frac{3}{4}} \\ -\ \mathbf{2\frac{1}{4}} \\ \hline \mathbf{2\frac{2}{4}} \end{array}$$

You can use a number line to visualize the problem.

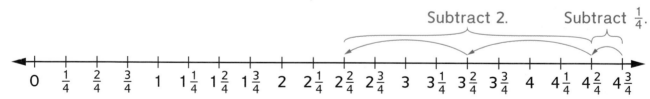

$4\frac{3}{4} - 2\frac{1}{4} = \mathbf{2\frac{2}{4}}$

$2\frac{2}{4}$ is close to the estimate of 3, so the answer makes sense.

When the fraction part you are taking away is greater than the fraction part of the starting number, you can decompose the mixed numbers to find equivalent mixed numbers or fractions to subtract. Think of trading wholes for groups of fractional parts.

Example

$2\frac{1}{6} - 1\frac{5}{6} = ?$

Estimate: $1\frac{5}{6}$ is close to 2, so the difference will be close to $2\frac{1}{6} - 2$, which equals $\frac{1}{6}$.

One way:

The fraction part to be subtracted is greater than the fraction part in the starting number. So rename $2\frac{1}{6}$ as an equivalent mixed number.

Show $2\frac{1}{6}$.

Break up 1 whole into 6 sixths to show that $2\frac{1}{6}$ is the same as 1 whole and 7 sixths, or $1\frac{7}{6}$.

Trade 1 whole for 6 sixths.

Rewrite the problem, then subtract $1\frac{5}{6}$.

$$2\frac{1}{6} \rightarrow \quad 1\frac{7}{6}$$
$$-1\frac{5}{6} \qquad -1\frac{5}{6}$$
$$\overline{\qquad} \qquad \overline{0\frac{2}{6}}$$

Take away $1\frac{5}{6}$. $\frac{2}{6}$ is left.

0 wholes and 2 sixths, or $\frac{2}{6}$, is the difference.

Another way:

Convert both mixed numbers into fractions greater than 1.

$2\frac{1}{6} = \frac{6}{6} + \frac{6}{6} + \frac{1}{6} = \frac{13}{6}$ 　　　 $1\frac{5}{6} = \frac{6}{6} + \frac{5}{6} = \frac{11}{6}$

Subtract the fractions.

$\frac{13}{6} - \frac{11}{6} = \frac{2}{6}$ 　

$2\frac{1}{6} - 1\frac{5}{6} = \frac{2}{6}$ The answer makes sense because $\frac{2}{6}$ is close to the estimate of $\frac{1}{6}$.

Adding Tenths and Hundredths

One way to add or subtract fractions with unlike denominators is to rename each fraction with like denominators, or a **common denominator.**

Example

$\frac{2}{10} + \frac{34}{100} = ?$

One way:

Change $\frac{2}{10}$ to an equivalent fraction with a denominator of 100.

$$\frac{(2 * 10)}{(10 * 10)} = \frac{20}{100}$$

Add the fractions with like denominators.

$$\frac{20}{100} + \frac{34}{100} = \frac{(20 + 34)}{100} = \frac{54}{100}$$

So, $\frac{2}{10} + \frac{34}{100} = \frac{54}{100}$.

Another way:

Use base-10 blocks to visualize the problem.

$\frac{2}{10} + \frac{34}{100} = ?$

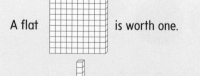

$$\frac{54}{100}$$

$$\frac{2}{10} + \frac{34}{100} = \frac{54}{100}$$

For the example on this page:

A flat [grid] is worth one.

A long [rod] is worth $\frac{1}{10}$, or 0.1.

A cube ⬦ is worth $\frac{1}{100}$, or 0.01.

Example

Rachel has 100 juice boxes for the school bake sale. $\frac{30}{100}$ are grape juice and $\frac{1}{10}$ are apple juice. What fraction of the juice boxes are either grape or apple juice?

Write a number sentence for the problem:

$$\frac{30}{100} + \frac{1}{10} = ?$$

The examples on this page and the next use shorthand pictures to represent base-10 blocks:

A flat □ is worth one.

A long | is worth $\frac{1}{10}$, or 0.1.

A cube ■ is worth $\frac{1}{100}$, or 0.01.

One way:

Rename $\frac{1}{10}$ as an equivalent fraction with a denominator of 100.

$$\frac{1}{10} = \frac{(1 * 10)}{(10 * 10)} = \frac{10}{100}$$

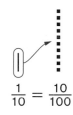

$$\frac{1}{10} = \frac{10}{100}$$

Rewrite the problem using the renamed fraction.

$$\frac{30}{100} + \frac{10}{100} = \frac{(30 + 10)}{100} = \frac{40}{100}$$

40 out of 100 juice boxes or $\frac{40}{100}$ of the juice boxes are either grape or apple juice.

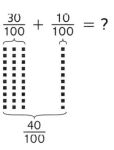

$$\frac{30}{100} + \frac{10}{100} = ?$$

$$\frac{40}{100}$$

Another way:

When the numerator is a multiple of 10, fractions with 100 in the denominator can also be expressed as a fraction with 10 in the denominator.

Rename $\frac{30}{100}$ as an equivalent fraction with 10 as the denominator.

$$\frac{30}{100} = \frac{3}{10}$$

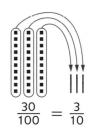

$$\frac{30}{100} = \frac{3}{10}$$

Use the new fraction to write the number sentence.

$$\frac{3}{10} + \frac{1}{10} = \frac{(3 + 1)}{10} = \frac{4}{10}$$

$\frac{4}{10}$ of Rachel's juice boxes are either grape or apple juice.

$$\frac{3}{10} + \frac{1}{10} = ?$$

$$\frac{4}{10}$$

Adding Mixed Numbers with Tenths and Hundredths

You can change the denominators of mixed numbers to add them.

Example

$3\frac{2}{10} + 1\frac{59}{100} = ?$

Write both mixed numbers with the same denominator in the fraction part. In this case, make the common denominator 100.

$3\frac{2}{10} = 3\frac{(2 * 10)}{(10 * 10)} = 3\frac{20}{100}$

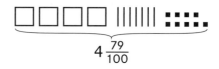

Decompose the mixed numbers into the sums of whole numbers and fractions.

$3\frac{20}{100} + 1\frac{59}{100} = 3 + \frac{20}{100} + 1 + \frac{59}{100}$

Combine the whole numbers and fractions.

Combine the blocks by type.

$3 + 1 = 4$

$\frac{20}{100} + \frac{59}{100} = \frac{(20 + 59)}{100} = \frac{79}{100}$

$4 + \frac{79}{100} = 4\frac{79}{100}$

$4\frac{79}{100}$

Or use fraction names:

3 *wholes* and 20 *hundredths* + 1 *whole* and 59 *hundredths* = 4 *wholes* and 79 *hundredths*.

Or replace each mixed number with an equivalent fraction:

$3\frac{20}{100} + 1\frac{59}{100} = \frac{320}{100} + \frac{159}{100} = \frac{479}{100}$

$3\frac{2}{10} + 1\frac{59}{100} = \mathbf{4\frac{79}{100}}$, or 4 *wholes* and 79 *hundredths*, or $\frac{479}{100}$.

Adding Decimals

You can think about decimals as fractions with a denominator of 10 or 100 to help you add.

> **Example**
>
> 0.6 + 0.33 = **?**
>
> **One way:**
>
> Rewrite the decimals as fractions.
>
> $0.6 = \frac{6}{10}$ $0.33 = \frac{33}{100}$
>
> Make sure both addends have the same denominator.
>
> $\frac{6}{10} = \frac{(6 * 10)}{(10 * 10)} = \frac{60}{100}$
>
> Rewrite the problem using fractions with like denominators.
>
> $\frac{60}{100} + \frac{33}{100} = \frac{(60 + 33)}{100} = \frac{93}{100}$
>
> Change the fraction back to a decimal.
>
> $\frac{93}{100} = \mathbf{0.93}$
>
> **Another way:**
>
> Think about each addend as a different color: 0.6 + 0.33 = ?
>
>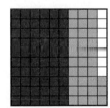
>
> Shade each decimal part in a different color on a hundredths grid, filling columns as you go.
> Count the total number of shaded columns.
> Each column represents 1 tenth. There are 6 red columns and 3 blue columns.
> That makes 9 tenths in all.
> Add the remaining 3 small blue squares, which represent 3 hundredths.
> There are 9 tenths (columns) and 3 hundredths (small squares) shaded on the grid.
> There are 93 hundredths (small squares) shaded on the grid.
>
> So, 0.6 + 0.33 = **0.93.**

> **Check Your Understanding**
>
> Add these fractions or decimals.
>
> **1.** $\frac{1}{10} + \frac{1}{100}$ **2.** $\frac{35}{100} + \frac{3}{10}$ **3.** $\frac{70}{100} + \frac{1}{10}$
>
> **4.** $2\frac{1}{10} + 5\frac{22}{100}$ **5.** 0.45 + 0.02
>
> Check your answers in the Answer Key.

Finding Fractions of a Set

Many problems with fractions involve finding a fraction of a number.

Example

There are 24 students in Ms. Dunning's class. $\frac{1}{3}$ of the students are participating in a school performance. How many students in Ms. Dunning's class are participating in the performance?

First identify the whole as 24 students.

Whole
24 students

To find the number of students participating in the performance, find $\frac{1}{3}$ of 24 students.

Model the problem using 24 counters. Each counter represents 1 student. Divide the counters into 3 equal groups, or thirds.

Each group has $\frac{1}{3}$ of the set of counters. There are 8 counters in each group.

So, $\frac{1}{3}$ of 24 counters is 8 counters.

That means that $\frac{1}{3}$ of 24 students is 8 students.
So, 8 of Ms. Dunning's students are participating in the school performance.

Example

A jacket that costs $46 is on sale for $\frac{1}{2}$ the regular price. What is the sale price?

To find the sale price, find $\frac{1}{2}$ of $46.

Model the problem using a rectangle to represent the whole cost. Divide the rectangle into two equal parts.

Divide 46 by 2 and label each half.

Half of $46 is $23. The sale price is $23.

Whole
$46

46

23	23

$46 \div 2 = 23$, so $\frac{1}{2}$ of 46 is 23.

Check Your Understanding

Solve.

Rita earned $20 raking lawns. She decided to use $\frac{1}{4}$ of her earnings for spending money, and save the rest. How much money will Rita have to spend?

Check your answers in the Answer Key.

Fractions as Multiples of Unit Fractions

Fractions with a numerator of 1, such as $\frac{1}{2}$, $\frac{1}{4}$, and $\frac{1}{10}$, are called **unit fractions.**
Unit fractions name 1 part of the whole. When other fractions are built from multiple parts of the whole, you can think of them as multiples of unit fractions.

Example

Rename $\frac{3}{4}$ as a multiple of a unit fraction.

Represent $\frac{3}{4}$ with fraction circle pieces. If the red circle is one whole, then each yellow piece represents one-fourth.

Whole
red circle

It takes 3 one-fourth pieces to make three-fourths. So, $\frac{3}{4}$ is the same as $3\left[\frac{1}{4}s\right]$.

You can rename $\frac{3}{4}$ as $\frac{1}{4} + \frac{1}{4} + \frac{1}{4}$, or as $3 * \frac{1}{4}$.

Example

Devin makes breakfast for his family. For each person, he makes a bowl of yogurt, granola, and berries. He uses $\frac{1}{4}$ cup of blueberries for each serving. How many cups of blueberries will Devin need to make 5 servings?

Model the problem by drawing fraction circles. If the circle represents 1 whole cup of blueberries, then the shaded part represents $\frac{1}{4}$ cup of blueberries.

For one serving, Devin needs $\frac{1}{4}$ cup of blueberries.

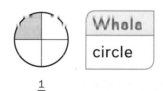

Whole
circle

$\frac{1}{4}$

For five servings, Devin needs five $\frac{1}{4}$ cups, or $\frac{5}{4}$ cups, of blueberries. $\frac{5}{4}$ cups is the same as 1 whole cup $\left(\frac{4}{4}\right)$ and $\frac{1}{4}$ cup more.

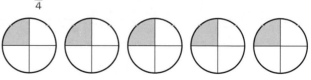

Addition equation: $\frac{1}{4} + \frac{1}{4} + \frac{1}{4} + \frac{1}{4} + \frac{1}{4} = \frac{5}{4}$

Multiplication equation: $5 * \frac{1}{4} = \frac{5}{4}$

Devin needs $\frac{5}{4}$ cups of blueberries to make five servings. This is the same as $1\frac{1}{4}$ cups of blueberries.

Example

Represent $\frac{3}{8}$ as a multiple of a unit fraction.

Locate $\frac{3}{8}$ on a number line. If the distance between 0 and 1 is divided into 8 equal parts, $\frac{3}{8}$ is 3 one-eighth jumps from 0 towards 1.

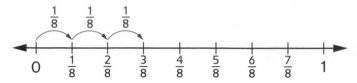

These 3 one-eighth jumps can be represented with addition and with multiplication.

Addition: $\frac{1}{8} + \frac{1}{8} + \frac{1}{8} = \frac{3}{8}$

Multiplication: $3 * \frac{1}{8} = \frac{3}{8}$

The number line and equations all show that three-eighths is the same as 3 copies of $\frac{1}{8}$.

Examples

Use what you know about unit fractions and multiples to solve each multiplication problem.

$9 * \frac{1}{2} = ?$ 9 times $\frac{1}{2}$ is the same as 9 halves, or $\frac{9}{2}$. $9 * \frac{1}{2} = \frac{9}{2}$

$4 * \frac{1}{5} = ?$ 4 times $\frac{1}{5}$ is the same as 4 fifths, or $\frac{4}{5}$. $4 * \frac{1}{5} = \frac{4}{5}$

$7 * \frac{1}{3} = ?$ 7 times $\frac{1}{3}$ is the same as 7 thirds, or $\frac{7}{3}$. $7 * \frac{1}{3} = \frac{7}{3}$

Check Your Understanding

1. Write an addition equation and a multiplication equation to describe the picture.

Whole
circle

2. Draw a picture to represent the following equations:

$\frac{1}{3} + \frac{1}{3} + \frac{1}{3} + \frac{1}{3} = \frac{4}{3}$ $4 * \frac{1}{3} = \frac{4}{3}$

Check your answers in the Answer Key.

Multiplying Fractions by Whole Numbers

There are several ways to think about multiplying a whole number and a fraction.

Using Fraction Circles

You can represent the multiplication problem with fraction circles. The whole number tells how many groups to make and the fraction tells what to put in each group.

> ### Example
>
> Use fraction circles to solve $4 * \frac{2}{3}$.
>
> Make 4 groups of $\frac{2}{3}$ with fraction circle pieces.
>
> There are 8 one-third pieces in all, so $4 * \frac{2}{3} = \frac{8}{3}$.
>
>
>
> | Whole |
> | --- |
> | red circle |

Drawing a Picture

You can draw pictures to multiply a whole number and a fraction.

> ### Example
>
> Use a picture to solve $4 * \frac{2}{3}$.
>
> Draw rectangles to represent wholes. Split each rectangle into three equal parts to show thirds.
>
> Shade 4 groups of $\frac{2}{3}$.
>
> In total, 8 one-third boxes are shaded, or $2\frac{2}{3}$ rectangles. So, $4 * \frac{2}{3} = \frac{8}{3}$, or $2\frac{2}{3}$.

Using a Number Line

You can think about counting up on a number line to multiply a whole number and a fraction. The whole number tells how many hops to make, and the fraction tells the length of each hop.

> ### Example
>
> Use a number line to solve $4 * \frac{2}{3}$.
>
> Sketch a number line that shows thirds.
>
> Start at 0 and count up four $\frac{2}{3}$-hops on the number line.
>
>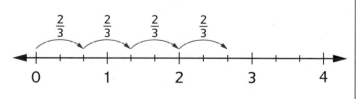
>
> In total, you traveled eight $\frac{1}{3}$-hops from 0. You end at $2\frac{2}{3}$. So, $4 * \frac{2}{3} = \frac{8}{3}$, or $2\frac{2}{3}$.

Number and Operations—Fractions

Using Mental Strategies

You can multiply a whole number and a fraction by mentally grouping the fractions.

> **Example**
>
> Use mental math to solve $4 * \frac{2}{3}$.
>
> Two $\frac{2}{3}$s are $\frac{4}{3}$, so four $\frac{2}{3}$s are $\frac{8}{3}$.
>
> $4 * \frac{2}{3} = \frac{8}{3}$

Using Repeated Addition

You can use repeated addition to multiply a fraction and a whole number.

> **Example**
>
> Use addition to solve $4 * \frac{2}{3}$.
>
> $4 * \frac{2}{3}$ is like making 4 copies of $\frac{2}{3}$. So, $4 * \frac{2}{3} = \frac{2}{3} + \frac{2}{3} + \frac{2}{3} + \frac{2}{3}$.
>
> Adding $\frac{2}{3}$ at a time, you get 2 thirds, 4 thirds, 6 thirds, and finally, 8 thirds.
>
> $\frac{2}{3} + \frac{2}{3} + \frac{2}{3} + \frac{2}{3} = \frac{8}{3}$, so $4 * \frac{2}{3} = \frac{8}{3}$.

Thinking about Multiples of Unit Fractions

You can use what you know about representing fractions as multiples of unit fractions to multiply any fraction and a whole number.

> **Example**
>
> Use what you know about multiples and unit fractions to solve $4 * \frac{2}{3}$.
>
> Rename the fraction as a multiple of a unit fraction. $\frac{2}{3}$ is the same as $2 * \frac{1}{3}$.
> Rewrite the multiplication problem.
>
> 4 groups of $\frac{2}{3}$ is the same as 4 groups of $\left(2 * \frac{1}{3}\right)$. $4 * \frac{2}{3} = 4 * \left(2 * \frac{1}{3}\right)$
>
> Multiply the whole numbers first: $4 * 2 = 8$ $= (4 * 2) * \frac{1}{3}$
>
> Then multiply the resulting whole number by the unit fraction. $= 8 * \frac{1}{3}$
>
> $8 * \frac{1}{3}$ is the same as eight-thirds, or $\frac{8}{3}$. $= \frac{8}{3}$
> So, $4 * \frac{2}{3} = \frac{8}{3}$.

> **Check Your Understanding**
>
> Use any method to solve these problems.
>
> **1.** $3 * \frac{4}{5}$ **2.** $2 * \frac{3}{4}$ **3.** $4 * \frac{5}{6}$ **4.** $5 * \frac{3}{8}$
>
> Check your answers in the Answer Key.

Multiplying Mixed Numbers by Whole Numbers

There are several ways to multiply a mixed number and a whole number.

Drawing a Picture

One way to multiply a mixed number and a whole number is to draw a picture.

Example

Solve $4 * 2\frac{3}{4}$.

Draw 4 copies of $2\frac{3}{4}$.

Find the total number of wholes: 4 rows of 2 wholes each makes 8 wholes.

Add the fractional parts: 4 rows of $\frac{3}{4}$ makes $\frac{12}{4}$, or 3 wholes.

Combine the total number of wholes and the sum of the fractional parts: $8 + 3 = 11$.

$4 * 2\frac{3}{4} = 11$

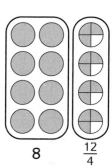

8     $\frac{12}{4}$

Using Repeated Addition

You can also use repeated addition.

Example

Solve $4 * 2\frac{3}{4}$.

Multiplying $4 * 2\frac{3}{4}$ is like making 4 copies of $2\frac{3}{4}$.

Break up each mixed number into a whole number and a fraction.

Group the whole numbers together and group the fractions together.

Add the wholes and add the fractions.

Rename the sum of the fractions as a mixed or whole number.

Add the sum of the wholes and the sum of the fractions.

$4 * 2\frac{3}{4} = 11$

$4 * 2\frac{3}{4} = 2\frac{3}{4} + 2\frac{3}{4} + 2\frac{3}{4} + 2\frac{3}{4}$

$= 2 + \frac{3}{4} + 2 + \frac{3}{4} + 2 + \frac{3}{4} + 2 + \frac{3}{4}$

$= (2 + 2 + 2 + 2) + \left(\frac{3}{4} + \frac{3}{4} + \frac{3}{4} + \frac{3}{4}\right)$

$= 8 + \frac{12}{4}$

$= 8 + 3$

$= 11$

Natural Measures and Standard Units

Systems of weights and measures have been used in many parts of the world since ancient times. People measured lengths and weights long before they had rulers and scales.

Ancient Measures of Weight

Shells and grains, such as wheat or rice, were often used as units of weight. For example, a small item might be said to weigh 300 grains of rice. Large weights were often compared to the load that could be carried by a man or pack animal.

Ancient Measures of Length

People used natural measures based on the human body to measure length and distance. Some of these units are shown below.

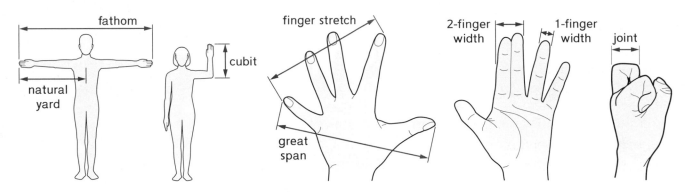

Standard Units of Length and Weight

Using shells and grains to measure weight is not exact. Even if the shells and grains are of the same type, they vary in size and weight.

Using body lengths to measure length is not exact either. Body measures depend upon the person who is doing the measuring. The problem is that different persons have hands and arms of different lengths.

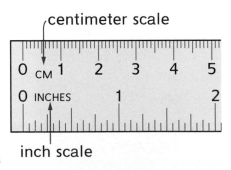

One way to solve this problem is to make **standard units** of length and weight. Most rulers are marked off using inches and centimeters as standard units. Bath scales are labeled using pounds and kilograms as standard units. Standard units never change and are the same for everyone. If two people measure the same object using standard units, their measurements will be the same or almost the same.

Did You Know?

The first metric standards were adopted in France in 1799. They were a standard meter and a kilogram bar.

The Metric System and the U.S. Customary System

The Metric System

About 200 years ago, a system of weights and measures called the **metric system** was developed. The metric system uses standard units for length, mass, and liquid volume. In the metric system:

- The **meter** is the standard unit for length. The symbol for a meter is **m.** A meter is about the width of a front door. Other commonly used metric units of length include the **millimeter, centimeter,** and **kilometer.**

- The **gram** is the standard unit for mass. The symbol for a gram is **g.** A large paper clip weighs about 1 gram. Another commonly used metric unit of mass is the **kilogram.**

- The **liter** is the standard unit for liquid volume or capacity. The symbol for a liter is **L.** Three regular-size canned drinks are about a liter. Another commonly used metric unit of liquid volume is the **milliliter.**

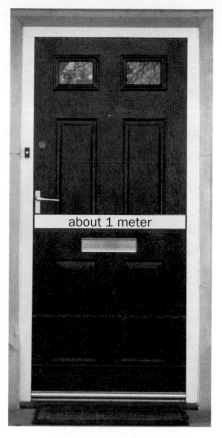
about 1 meter

Scientists almost always use the metric system for measurement. It is easy to use because it is a base-10 system. Larger and smaller units are defined by multiplying or dividing the standard units (given above) by multiples of ten: 10, 100, 1,000, and so on.

Note The U.S. customary system is not based on multiples of ten. This makes it more difficult to use than the metric system. For example, to change inches to yards, you must know that 36 inches equals 1 yard.

The U.S. Customary System

The metric system is used in most countries around the world. In the United States, however, the **U.S. customary system** is used for everyday purposes. In the U.S. customary system:

- The **yard** is the standard unit for length. The symbol for a yard is **yd.** A yard is about the distance from the center of an adult's chest to the tip of a finger. Other commonly used U.S. customary units of length include the **inch, foot,** and **mile.**

- The **pound** is the standard unit for weight. The symbol for a pound is **lb.** A package of 4 sticks of butter is about 1 pound. Other commonly used U.S. customary units of weight include the **ounce** and **ton.**

- The **gallon** is the standard unit for liquid volume or capacity. The symbol for a gallon is **gal.** A gallon of milk is commonly used as an example of a gallon. Other commonly used U.S. customary units of capacity include the **cup, pint,** and **quart.**

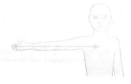
About 1 yard

Chris Rose/E+/Getty Images

Length: Metric System

Length is the measure of the distance between two points. In the metric system, the basic unit for length or distance is the meter (m). Commonly used metric units of measure for length include millimeters (mm), centimeters (cm), and kilometers (km). Length in the metric system is usually measured with a meterstick, the centimeter side of a ruler, or a metric tape measure.

Examples of Tools for Measuring Length

meterstick

tape measure

Part of a centimeter ruler is shown here. The ruler shows centimeters divided into millimeters.

1 centimeter   1 millimeter

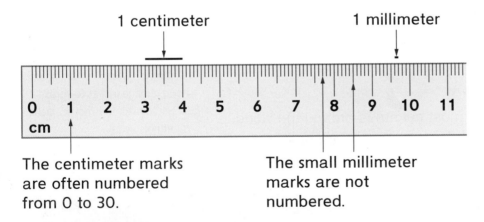

The centimeter marks are often numbered from 0 to 30.

The small millimeter marks are not numbered.

Check Your Understanding

1. Measure the line below using the metric side of a ruler. Meaure to the nearest tenth of a centimeter (cm).

   _____

2. Draw a line that is 4.3 centimeters long.

   Check your answers in the Answer Key.

Personal References

You can estimate lengths by using the lengths of common objects and distances that you know. These are called personal references. Some examples of personal references for metric units of length are given below.

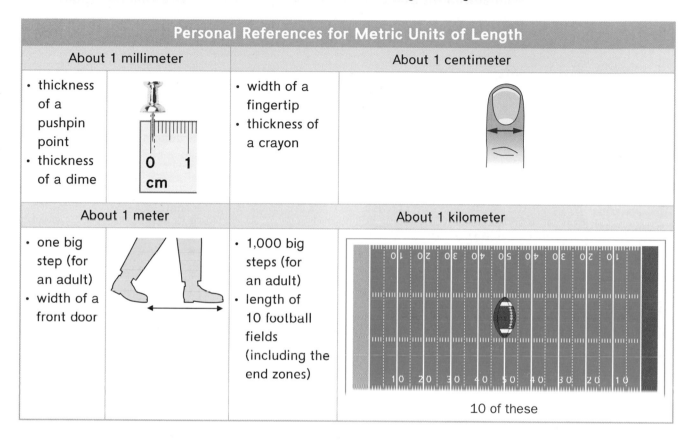

Personal References for Metric Units of Length	
About 1 millimeter	**About 1 centimeter**
• thickness of a pushpin point • thickness of a dime	• width of a fingertip • thickness of a crayon
About 1 meter	**About 1 kilometer**
• one big step (for an adult) • width of a front door	• 1,000 big steps (for an adult) • length of 10 football fields (including the end zones) 10 of these

The table below shows how some different units of length in the metric system compare.

Metric System
Units of Length
1 kilometer (km) = 1,000 meters (m)
1 meter = 100 centimeters (cm)
1 meter = 1,000 millimeters (mm)

Converting Metric Units of Length

Measurement scales and two-column tables are two types of representations that can help you convert one metric unit of length to another.

Measurement Scale

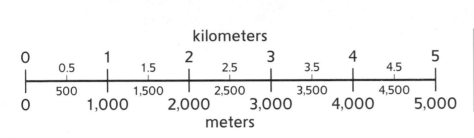

Two-Column Table

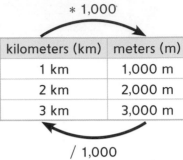

kilometers (km)	meters (m)
1 km	1,000 m
2 km	2,000 m
3 km	3,000 m

Examples

Use the measurement scale or table above to convert a given length to a different unit. *Think:* Does the answer make sense?

$$4 \text{ kilometers} = \underline{\hspace{2cm}} \text{ meters}$$

Solve Using the Measurement Scale

Using the measurement scale with kilometers and meters, locate the side of the scale that represents kilometers. Then locate 4 kilometers. Next find the matching measurement in meters on the other side of the scale.

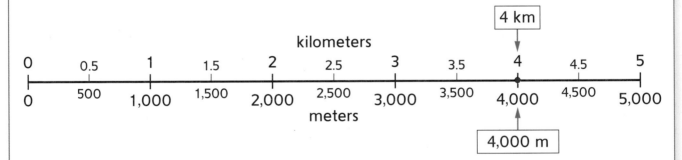

Solve Using the Two-Column Table

According to the two-column table showing kilometers and meters, multiply the number of kilometers by 1,000 meters in one kilometer to find the equivalent number of meters.

$$4 \text{ km} = 4 * 1,000 \text{ m} = \textbf{4,000 m}$$

This involves converting a larger unit of length to a smaller unit of length. That means it takes a greater number of smaller units to represent the same length. The answer makes sense.

Measurement Scales

centimeters

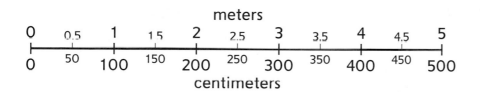

millimeters

meters

centimeters

Two-Column Tables

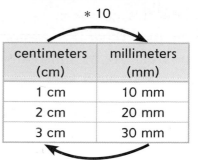

* 10

centimeters (cm)	millimeters (mm)
1 cm	10 mm
2 cm	20 mm
3 cm	30 mm

/ 10

* 100

meters (m)	centimeters (cm)
1 m	100 cm
2 m	200 cm
3 m	300 cm

/ 100

Check Your Understanding

1. Complete the two-column tables to convert metric lengths to a different unit.

meters (m)	centimeters (cm)
2 m	
3.5 m	
6 m	

kilometers (km)	meters (m)
4 km	
1.5 km	
6 km	

2. What metric unit of length would you use to measure the height of a house?

Check your answers in the Answer Key.

Length: U.S. Customary System

In the U.S. customary system, the basic unit for length or distance is the yard (yd). Other U.S. customary units of measure for length include inches (in.), feet (ft), and miles (mi). Length in the U.S. customary system is usually measured with a ruler, tape measure, or yardstick.

Examples of Tools for Measuring Length

ruler

tape measure

On rulers, inches are usually divided into halves, quarters, eighths, and sixteenths. There are usually different-size marks to show different fractions of an inch.

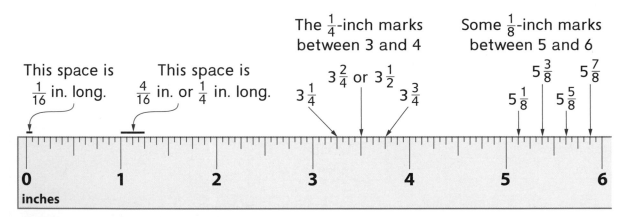

This space is $\frac{1}{16}$ in. long.

This space is $\frac{4}{16}$ in. or $\frac{1}{4}$ in. long.

The $\frac{1}{4}$-inch marks between 3 and 4

$3\frac{2}{4}$ or $3\frac{1}{2}$

$3\frac{1}{4}$ $3\frac{3}{4}$

Some $\frac{1}{8}$-inch marks between 5 and 6

$5\frac{3}{8}$ $5\frac{7}{8}$

$5\frac{1}{8}$ $5\frac{5}{8}$

Check Your Understanding

1. Measure the line below to the nearest half-inch.

_____

2. Draw a line that is $2\frac{1}{4}$ inches long.

Check your answers in the Answer Key.

Personal References

You can estimate lengths by using the lengths of common objects and distances that you know. These are called personal references. Some examples of personal references for U.S. customary units of length are given below.

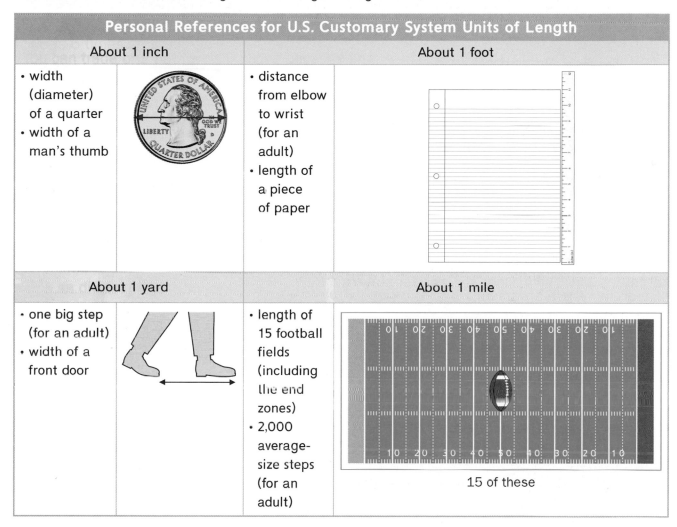

Personal References for U.S. Customary System Units of Length	
About 1 inch	**About 1 foot**
• width (diameter) of a quarter • width of a man's thumb	• distance from elbow to wrist (for an adult) • length of a piece of paper
About 1 yard	**About 1 mile**
• one big step (for an adult) • width of a front door	• length of 15 football fields (including the end zones) • 2,000 average-size steps (for an adult) 15 of these

The table below shows how some different units of length in the U.S. customary system compare.

U.S. Customary System
Units of Length
1 mile (mi) = 1,760 yards (yd)
1 mile = 5,280 feet (ft)
1 yard = 3 feet
1 yard = 36 inches (in.)
1 foot = 12 inches

Converting U.S. Customary System Units of Length

Measurement scales and two-column tables are two types of representations that can help you convert one U.S. customary unit of length to another U.S. customary unit of length.

Measurement Scale

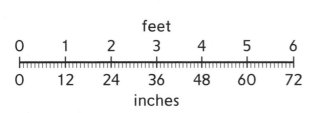

Two-Column Table

	* 12	
feet (ft)	inches (in.)
1 ft	12 in.
2 ft	24 in.
3 ft	36 in.
	/ 12	

Examples

Use the measurement scale or table above to convert a given length to a different unit. *Think: Does the answer make sense?*

$$5 \text{ feet} = \underline{\hspace{2cm}} \text{ inches}$$

Solve Using the Measurement Scale

Using the measurement scale with feet and inches, locate the side of the scale that represents feet. Then locate 5 feet. Next find the matching measurement in inches on the other side of the scale.

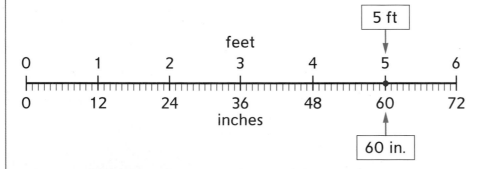

Solve Using the Two-Column Table

According to the two-column table showing feet and inches, multiply the number of feet by 12 inches in one foot to find the equivalent number of inches.

$$5 \text{ ft} = 5 * 12 \text{ in.} = \mathbf{60 \text{ in.}}$$

This involves converting a larger unit of length to a smaller unit of length. That means it takes a greater number of smaller units to represent the same length. The answer makes sense.

Measurement Scales

yards

0   1   2   3

0   1   2   3   4   5   6   7   8   9

feet

yards

0   1   2   3

0   12   24   36   48   60   72   84   96   108

inches

Two-Column Tables

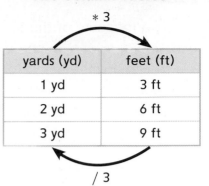

* 3

yards (yd)	feet (ft)
1 yd	3 ft
2 yd	6 ft
3 yd	9 ft

/ 3

* 36

yards (yd)	inches (in.)
1 yd	36 in.
2 yd	72 in.
3 yd	108 in.

/ 36

Check Your Understanding

1. Complete these two-column tables to convert lengths in the U.S. customary system to a different unit.

yards (yd)	feet (ft)
2 yd	
6 yd	
10 yd	

feet (ft)	inches (in.)
3 ft	
$4\frac{1}{2}$ ft	
12 ft	

2. What U.S. customary unit of length would you use to measure the length of a car?

Check your answers in the Answer Key.

Mass

Mass is a measure of the amount of matter (solid, liquid, or gas) in an object. The gram (g) is the basic metric unit for mass. Another metric unit of measure for mass is the kilogram (kg). One tool that can be used for measuring mass is a pan balance.

pan balance

Personal References

You can estimate masses by using the masses of common objects that you know. These are called personal references.

| 1 g | 5 g | 10 g | 20 g | 100 g | 1,000 g |

masses

Some examples of personal references for mass are given below.

Personal References for Metric Units of Mass	
About 1 gram	About 1 kilogram
• paper dollar bill • large paper clip	• wooden bat • 200 U.S. nickels

Converting Metric Units of Mass

Measurement scales and two-column tables are two types of representations that can help you convert one metric unit of mass to another metric unit of mass.

Note Mass is measured in kilograms and grams in the metric system and in pounds and ounces in the U.S. customary system.

Measurement Scale

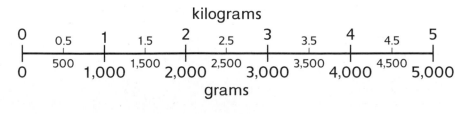

kilograms

| 0 | 0.5 | 1 | 1.5 | 2 | 2.5 | 3 | 3.5 | 4 | 4.5 | 5 |

| 0 | 500 | 1,000 | 1,500 | 2,000 | 2,500 | 3,000 | 3,500 | 4,000 | 4,500 | 5,000 |

grams

Two-Column Table

＊1,000

kilograms (kg)	grams (g)
1 kg	1,000 g
2 kg	2,000 g
3 kg	3,000 g

/ 1,000

Examples

Use the measurement scale or the table on the previous page to convert a given mass of an object to a different unit. *Think: Does the answer make sense?*

5 kilograms = _____ grams

Solve Using the Measurement Scale

Using the measurement scale with kilograms and grams, locate the side of the scale that represents kilograms. Then locate 5 kilograms. Next find the matching measurement in grams on the other side of the scale.

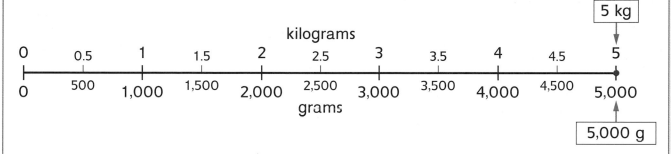

Solve Using the Two-Column Table

According to the two-column table showing kilograms and grams, multiply the number of kilograms by 1,000 grams in one kilogram to find the equivalent number of grams.

5 kg = 5 * 1,000 g = **5,000 g**

This involves converting a larger unit of mass to a smaller unit of mass. That means it takes a greater number of smaller units to represent the same mass. The answer makes sense.

Check Your Understanding

1. Complete the two-column table to convert the number of kiograms to grams.

kilograms (kg)	grams (g)
3 kg	
2.5 kg	
19 kg	

2. What metric unit of mass would you use to measure the mass of a cat? a baseball?

Check your answers in the Answer Key.

Weight

In everyday life, many people use the words *mass* and *weight* to mean the same thing, but scientifically, there is a difference. Mass is a measure of the amount of matter (solid, liquid, and gas) in an object. **Weight** is a measure of the heaviness of an object. Weight depends both on an object's mass and on the pull of gravity.

On the moon, an astronaut's body has the same amount of matter, or mass, as on Earth. However, the astronaut's weight is much less on the moon than on Earth because the moon has less gravitational pull.

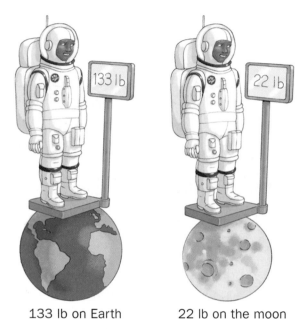

133 lb on Earth   22 lb on the moon

In everyday life in the United States, we use ounces (oz), pounds (lb), and tons (T) to measure weight. You can use tools such as bath scales and spring scales to measure weight.

bath scale

spring scale

Personal References

You can estimate weights by using the weights of common objects that you know. These are called personal references. Some examples of personal references for weight are given below.

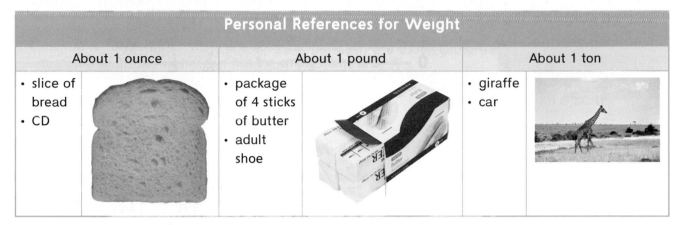

Personal References for Weight		
About 1 ounce	About 1 pound	About 1 ton
• slice of bread • CD	• package of 4 sticks of butter • adult shoe	• giraffe • car

Converting Units of Weight

Measurement scales and two-column tables are two types of representations that can help you convert one unit of weight to another unit of weight.

Measurement Scales

Two-Column Tables

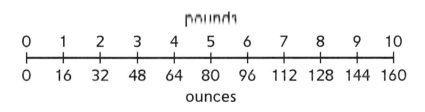

pounds

| 0 | 1 | 2 | 3 | 4 | 5 | 6 | 7 | 8 | 9 | 10 |

| 0 | 16 | 32 | 48 | 64 | 80 | 96 | 112 | 128 | 144 | 160 |

ounces

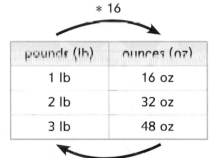

* 16

pounds (lb)	ounces (oz)
1 lb	16 oz
2 lb	32 oz
3 lb	48 oz

/ 16

* 2,000

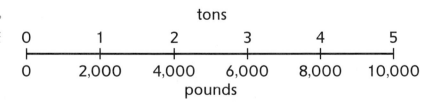

tons

| 0 | 1 | 2 | 3 | 4 | 5 |

| 0 | 2,000 | 4,000 | 6,000 | 8,000 | 10,000 |

pounds

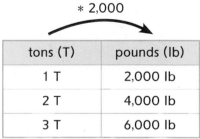

tons (T)	pounds (lb)
1 T	2,000 lb
2 T	4,000 lb
3 T	6,000 lb

/ 2,000

Examples

Use the measurement scales or tables on the previous page to convert a given weight to a different unit. *Think:* Does the answer make sense?

6 pounds = _____ ounces

Solve Using the Measurement Scale

Using the measurement scale with pounds and ounces, locate the side of the scale that represents pounds. Then locate 6 pounds. Next find the matching measurement in ounces on the other side of the scale.

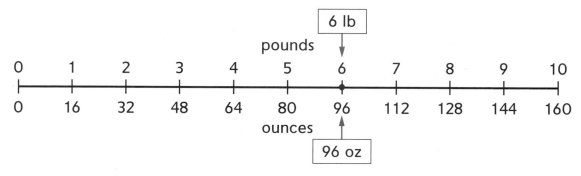

Solve Using the Two-Column Table

According to the two-column table showing pounds (lb) and ounces (oz), multiply the number of pounds by 16 ounces in one pound to find the equivalent number of ounces.

6 lb = 6 * 16 oz = **96 oz**

This involves converting a larger unit of weight to a smaller unit of weight. That means it takes a greater number of smaller units to represent the same weight. The answer makes sense.

Check Your Understanding

1. Complete the two-column tables to convert weight to a different unit.

pounds (lb)	ounces (oz)
2 lb	
7 lb	
15 lb	

tons (T)	pounds (lb)
5 T	
8 T	
20 T	

2. What unit of measure would you use to measure the weight of

   a sheet of paper?

   a blue whale?

   a watermelon?

Check your answers in the Answer Key.

Liquid Volume: Metric System

Liquid volume is a measure of the amount of space taken up by a liquid. In the metric system, units for liquid volume include milliliters (mL) and liters (L). You can use tools such as graduated cylinders and beakers to measure liquid volume.

100 mL graduated cylinder

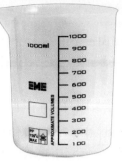

1,000 mL beaker, or 1 L beaker

Example

What is the volume of water in the beaker?

The **volume** of water in the beaker is 75 mL.

> **Note Capacity** is the amount a container can hold. The capacity of the beaker in the example is 100 mL, and the volume of water in the beaker is 75 mL.

Personal References

You can estimate liquid volume by using common objects that you know as personal references. Examples of personal references for metric units of liquid volume are given below.

Personal References for Metric Units of Liquid Volume	
About 1 milliliter	About 1 liter
• 20 drops of water • the amount of water that could fit in a centimeter cube* *Note that 1 mL is defined as 1 cu cm, so 1 cu cm of water is exactly 1 mL. 20 drops of water	• 3 regular-size canned drinks • 1-liter bottle 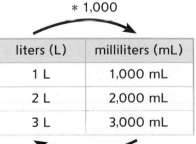

Converting Metric Units of Liquid Volume

Measurement scales and two-column tables are two types of representations that can help you convert a metric unit of liquid volume to another metric unit of liquid volume.

Measurement Scale

```
                    liters
0        1        2        3        4        5
|---|---|---|---|---|---|---|---|---|---|
    0.5      1.5      2.5      3.5      4.5
    500     1,500    2,500    3,500    4,500
0       1,000    2,000    3,000    4,000    5,000
                 milliliters
```

Two-Column Table

* 1,000

liters (L)	milliliters (mL)
1 L	1,000 mL
2 L	2,000 mL
3 L	3,000 mL

/ 1,000

one hundred ninety-three

**SRB
193**

Examples

Use the measurement scale or table on the previous page to convert a given liquid measure to a different unit. *Think:* Does the answer make sense?

2 liters = _____ milliliters

Solve Using the Measurement Scale

Using the measurement scale with liters and milliliters, locate the side of the scale that represents liters. Then locate 2 liters. Next find the matching measurement in milliliters on the other side of the scale.

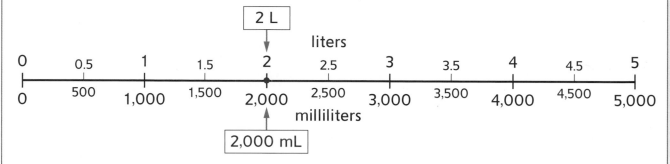

Solve Using the Two-Column Table

According to the two-column table showing liters and milliliters, multiply the number of liters by 1,000 milliliters in one liter to find the equivalent number of milliliters.

2 L = 2 * 1,000 mL = **2,000 mL**

This involves converting a larger unit of liquid volume to a smaller unit of volume. That means it takes a greater number of smaller units to represent the same volume. The answer makes sense.

Check Your Understanding

1. Complete the two-column table to convert liters to milliliters.

liters (L)	milliliters (mL)
2	
7	
39	

2. What unit of measure would you use to measure the volume of

water in a swimming pool?

soup in a bowl?

water in a kitchen sink?

Check your answers in the Answer Key.

Liquid Volume: U.S. Customary System

Liquid volume is a measure of the amount of space taken up by a liquid. In the U.S. customary system, units for liquid volume include cups (c), pints (pt), quarts (qt), and gallons (gal).

Examples of Tools for Measuring Liquid Volumes

measuring spoons

measuring cup

5-gallon bucket

Tools such as measuring spoons and cups can also be used to measure the volume of things besides liquids, such as flour and sugar.

Example

What is the volume of water in the measuring cup?

The amount of water measures $2\frac{1}{2}$ cups.

Personal References

You can estimate liquid volume by using common objects that you know. These are called personal references. Some examples of personal references for U.S. customary units of volume are given below.

Personal References for U.S. Customary Units of Liquid Volume			
About 1 cup	About 1 pint	About 1 quart	About 1 gallon
cup	pint	quart	gallon

The table at the right shows how some different units of liquid volume in the U.S. customary system compare.

Converting U.S. Customary Units of Liquid Volume

Measurement scales and two-column tables are two types of representations that can help you convert one U.S. customary unit of liquid volume to another.

U.S. Customary System
Units of Liquid Volume
1 gallon (gal) = 4 quarts (qt)
1 quart = 2 pints (pt)
1 pint = 2 cups (c)
1 cup = 8 fluid ounces (fl oz)

Measurement Scale

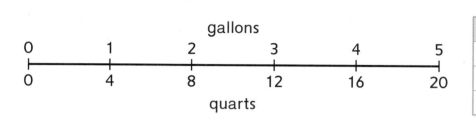

Two-Column Table

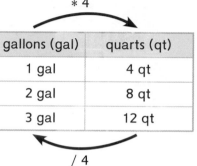

Examples

Use the measurement scale or table above to convert a given volume to a different unit. *Think:* Does the answer make sense?

$$3 \text{ gallons} = \underline{\hspace{2cm}} \text{ quarts}$$

Solve Using the Measurement Scale

Using the measurement scale, locate 3 gallons on the side of the scale that represents gallons. Then find the matching measurement in quarts on the other side of the scale.

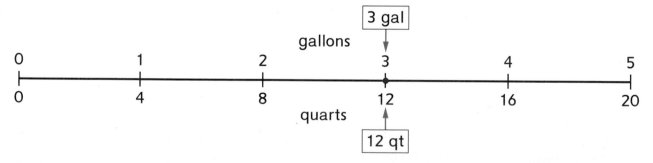

Solve Using the Two-Column Table

According to the two-column table showing gallons and quarts, multiply the number of gallons by 4 quarts in one gallon to find the equivalent number of quarts.

$$3 \text{ gal} = 3 * 4 \text{ qt} = \textbf{12 qt}$$

This involves converting a larger unit to a smaller unit. That means it takes a greater number of smaller units to represent the same volume. The answer makes sense.

Measurement Scales

pints

0	1	2	3	4	5
0	2	4	6	8	10

cups

quarts

0	1	2	3	4	5
0	2	4	6	8	10

pints

Two-Column Tables

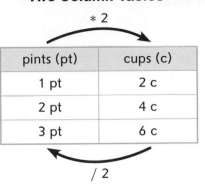

pints (pt)	cups (c)
1 pt	2 c
2 pt	4 c
3 pt	6 c

/ 2

* 2

quarts (qt)	pints (pt)
1 qt	2 pt
2 qt	4 pt
3 qt	6 pt

/ 2

Check Your Understanding

1. Complete the two-column tables to convert liquid volumes to a different unit.

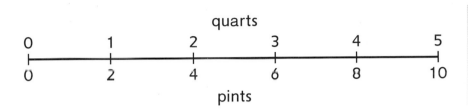

pints (pt)	cups (c)
3 pt	
5 pt	
10 pt	

quarts (qt)	pints (pt)
1 qt	
$4\frac{1}{2}$ qt	
12 qt	

gallons (gal)	quarts (qt)
2 gal	
4 gal	
25 gal	

2. What unit of measure would you use to measure the volume of

ice cream in a carton?

soup in a bowl?

water in a fish tank?

Check your answers in the Answer Key.

Time

We use time in two ways:

1. to tell when something happens, and

2. to tell how long something takes or lasts.

You can use analog clocks and digital clocks to show time.

Units of Time
1 century = 100 years (yr)
1 decade = 10 years
1 year = 12 months (mo)
1 year = 52 weeks (w) (plus one or two days)
1 year = 365 days (d) (366 days in a leap year)
1 month = 28, 29, 30, or 31 days
1 week = 7 days
1 day = 24 hours (hr)
1 hour = 60 minutes (min)
1 minute = 60 seconds (sec)

analog clock

digital clock

Units of time include seconds (sec), minutes (min), hours (hr), days (d), weeks (w), months (mo), and years (yr).

The table above shows how some different units of time compare.

Converting Units of Time

The measurement scale and two-column table below are two types of representations that can help you convert hours to minutes.

Measurement Scale

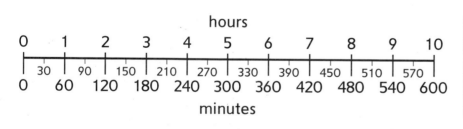

Two-Column Table

	* 60	
hours (hr)	minutes (min)
1 hr	60 min
2 hr	120 min
3 hr	180 min
	/ 60	

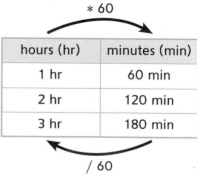

Examples

Use the measurement scales or tables on the previous page to convert a given time to a different unit. *Think:* Does the answer make sense?

8 hours = _____ minutes

Solve Using the Measurement Scale

Using the measurement scale with hours and minutes, locate the side of the scale that represents hours. Then locate 8 hours. Next find the matching measurement in minutes on the other side of the scale.

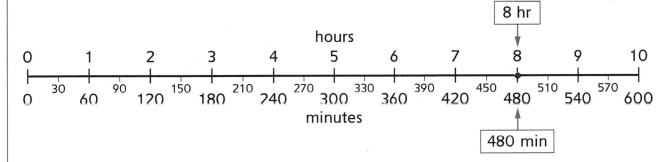

Solve Using the Two-Column Table

According to the two-column table showing hours and minutes, multiply the number of hours by 60 minutes in one hour to find the equivalent number of minutes.

8 hr = 8 * 60 min = **480 min**

This involves converting a larger unit of time to a smaller unit of time. That means it takes a greater number of smaller units to represent the same amount of time. The answer makes sense.

Check Your Understanding

Complete the two-column tables to show equal times using a different unit.

minutes	seconds
4 min	
9 min	
13 min	

hours	minutes
3 hr	
$7\frac{1}{2}$ hr	
20 hr	

Check your answers in the Answer Key.

Perimeter

Sometimes we want to know the distance around a shape. **Perimeter** is the length of the boundary of a closed shape. To measure perimeter, we use units of length such as inches, centimeters, meters, or miles.

To find the perimeter of any polygon, add the length of each of its sides. Remember to name the unit of length used to measure the polygon's sides.

A concrete wall creates a boundary around the pond. The length of this boundary is the perimeter of the pond.

Example

Find the perimeter of polygon *ABCDE*.

3 m + 8 m + 5 m + 4 m + 10 m = 30 m

The perimeter is 30 meters.

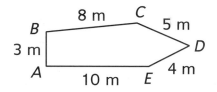

A rule, or **formula,** can be used for calculating the perimeter of some polygons. To find the perimeter of a rectangle, you can use any formula shown below. In each formula, *p* is the perimeter, *l* is the length of the rectangle, and *w* is the width of the rectangle.

Examples

Find the perimeter of the rectangle.

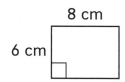

length (*l*) = 8 cm
width (*w*) = 6 cm

Add the measures of the four sides:

$$p = l + l + w + w$$

perimeter (*p*) = 8 cm + 8 cm + 6 cm + 6 cm = 28 cm

Add the width and length and double the sum:

$$p = 2 * (l + w)$$

perimeter (*p*) = 2 * (8 cm + 6 cm)
perimeter (*p*) = 2 * 14 cm
perimeter (*p*) = 28 cm

Double the length, double the width, and then add:

$$p = 2l + 2w$$

perimeter (*p*) = (2 * 8 cm) + (2 * 6 cm)
perimeter (*p*) = 16 cm + 12 cm
perimeter (*p*) = 28 cm

The perimeter of the rectangle is 28 centimeters.

Squares are special rectangles that have equal-length sides. You can use any of the perimeter formulas for rectangles to find the perimeter of a square. Because every side of a square is the same length, you can also use the formulas below to calculate the perimeter of a square. In each formula, p is the perimeter, and s is the length of a side of the square.

Examples

Find the perimeter of the square.

5 ft

All 4 sides have the same length.

The perimeter of the square is 20 feet.

$$p = s + s + s + s$$

length of a side (s) = 5 ft

perimeter (p) = 5 ft + 5 ft + 5 ft + 5ft = 20 ft

$$p = 4 * s$$

length of a side (s) = 5 ft

perimeter (p) = 4 * 5 ft = 20 ft

Check Your Understanding

1. Find the perimeter of the square.

16 mm

2. Find the perimeter of the rectangle.

12 m

$6\frac{1}{2}$ m

3. Measure the sides of a piece of paper to the nearest half-inch. What is the perimeter of the paper?

Check your answers in the Answer Key.

Area

Sometimes we want to know the amount of surface inside a shape. The amount of surface inside the boundary of a closed shape is called its **area.**

The images at the right represent two different backyards. Carla's backyard has a larger area than Greg's backyard. Greg's backyard could be placed inside of Carla's backyard with space left over.

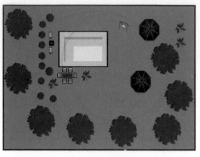

Carla's backyard   Greg's backyard

Area is measured in square units. You can find the area of a shape by counting the number of squares of a certain size that completely cover the inside of the shape. The squares must cover the entire inside of the shape and must not overlap, have any gaps, or cover any surface outside of the shape.

Example

What is the area of the hexagon?

The hexagon on the right is covered by squares that measure one unit on each side. Each square is called a **square unit (unit²).**

Five of the squares cover the hexagon.

The area of the hexagon is 5 square units.

This is written as 5 sq units or 5 units².

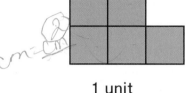

1 unit

1 square unit

1 unit

Reminder: Be careful not to confuse the area of a shape with its perimeter.

Area	**Perimeter**
Area is the amount of **surface inside the boundary** of a closed shape.	Perimeter is the **length of the boundary** of a closed shape.
Area is measured in square units such as square inches, square feet, square centimeters, square meters, and square miles.	Perimeter is measured in units such as inches, feet, centimeters, meters, and miles.
area of the purple space inside the square: 1 sq in.	perimeter of the square shown in red: 4 in.

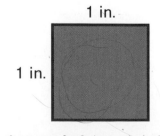

1 in.

1 in.

1 square inch (actual size)

Sometimes it is easier to find the area of the space inside a shape if you count groups of units. Smaller units can be grouped together into larger **composite units.** In the rectangle at the right, five squares with 1-foot sides are grouped together to form a composite unit that has an area of 5 square feet. The composite unit is shaded in red. It is one horizontal row of squares inside the rectangle.

Each square is 1 square foot.

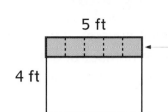

This composite unit has an area of 5 square feet.

You can find the area of a shape by adding the number of times you repeat the composite units as you cover the inside of the shape.

Example

Use the shaded composite unit to find the area of the rectangle.

Each square is 1 square foot. The composite unit has an area of 5 square feet.

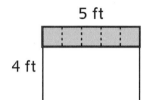

- There are 4 rows of squares, with 5 squares in each row.

- The composite unit repeats 4 times as it covers the entire surface of the rectangle.

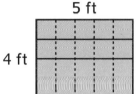

5 squares in the first row
10 squares in the first 2 rows
15 squares in the first 3 rows
20 squares in all 4 rows

- Skip count by 5 to count all of the squares in the rectangle: 5, 10, 15, 20.

The area of the rectangle is 20 square feet.

Check Your Understanding

Use the shaded composite unit to find the area of the rectangles below.

1. 3 cm

2 cm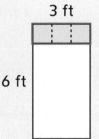

2. 3 ft

6 ft

3. 6 units

8 units

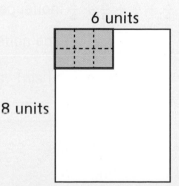

Check your answers in the Answer Key.

Area of a Rectangle

When you tile a rectangle with unit squares, the squares are arranged in rows. Each row contains the same number of squares. Imagine sweeping or rolling across the surface of a shape with a row. You can find the area of a rectangle by taking the number of squares in each row and multiplying the number of times you repeat each row as you cover the inside of a shape.

Example

Find the area of the rectangle.

There are 5 squares in each row and 3 rows in the rectangle. There are 15 squares.

Area = 15 square units

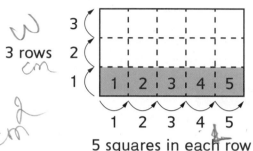

3 rows

5 squares in each row

The number of squares in one row is the **length** of the rectangle, and the number of rows in the rectangle is its **width.** To find the area of a rectangle, you can use the following formula:

area	=	(the number of squares in one row)	*	(the number of rows)
area	=	length	*	width
A	=	l	*	w

Examples

Find the area of each rectangle.
Use the formula $A = l * w$.

4 units

3 units

- number of squares per row: 4
  length (l) = 4 units
- number of rows: 3
  width (w) = 3 units
- area (A) = 4 units * 3 units = 12 sq units

The area of the rectangle is 12 square units.

Use the formula $A = l * w$.

3 mi

7 mi

- length (l) = 3 mi
- width (w) = 7 mi
- area (A) = 3 mi * 7 mi = 21 sq mi

The area of the rectangle is 21 square miles.

The shapes below are called **rectilinear figures** because their sides are all line segments and the inside and outside corners of each shape are all right angles. They are not rectangles because they have more than four sides.

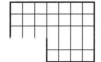 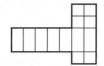

You cannot find the areas of these figures by multiplying the number of squares in each row by the number of rows. This is because the rows do not all have the same number of squares. You can find the area of a rectilinear figure by **decomposing,** or separating, it into non-overlapping rectangles, finding the area of each rectangle, and then adding the areas of those rectangles.

Note You can find other ways to decompose the shape in the example below into two or more rectangles. Then follow the same steps.

Example

Find the area of this rectilinear shape.
Each square is 1 square inch.

Step 1 Decompose the shape into two non-overlapping rectangles.

Step 2 Find the area of each rectangle. Use the formula $A = l * w$.

   area of blue rectangle = 3 inches * 3 inches = 9 square inches

   area of red rectangle = 4 inches * 5 inches = 20 square inches

Step 3 Add the areas of the two rectangles to find the area of the rectilinear shape.

   area of shape = area of blue rectangle + area of red rectangle

   = 9 square inches + 20 square inches

   = 29 square inches

The area of the shape is 29 square inches.

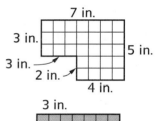
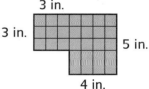

You could also look at the rectilinear shape in the example on the previous page as a large rectangle with a corner cut out. Another way to find the area of the rectilinear shape is to first find the area of the large rectangle, and then subtract the area of the small corner rectangle.

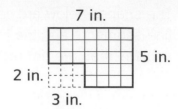

area of rectilinear shape	=	area of large rectangle	−	area of small corner rectangle
	=	(7 inches * 5 inches)	−	(3 inches * 2 inches)
	=	35 square inches	−	6 square inches
	=		29 square inches	

The area of the rectilinear shape is 29 square inches.

This checks the previous solution in the example on page 205.

Check Your Understanding

Find the area of the following figures. Include the unit in your answers.

1.   2 m

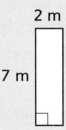

7 m

2.   6 in.

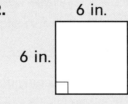

6 in.

3. Draw the shape in the example on page 205 on grid paper or make a sketch. Decompose it into different rectangles than those shown in the example. Find the sum of their areas. Did you find the same area as the area in the example?

2 cm

2 cm   5 cm

5 cm

4. Copy the figure at the right onto grid paper or make a sketch. Find the area of the figure.

Check your answers in the Answer Key.

Standard Units for Measuring Angles

When you describe length with numbers, you use standard units such as feet or meters. When you describe the size of angles with numbers, you can use a standard unit called a **degree.** Degrees are marked with a small raised circle (°).

If you divide a circle into 360 equal angles with their vertices at the center of the circle, each of those angles measures 1 degree. So 360 angles that each measure 1 degree and do not overlap will fill a circle.

Think of one ray of an angle rotating away from the other ray on a circle. Start both rays at the 0° mark. Rotate one ray clockwise around the circle (in the direction of the red arrows shown below). The degree numbers written on the circle show how many marks on the circle the ray has rotated away from the other ray. When the ray rotates clockwise $\frac{1}{4}$ of the full circle, it has rotated 90 degrees (360° / 4 = 90°).

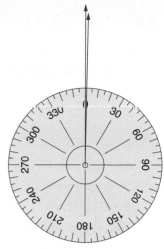

an angle that measures 1°

Examples

The measure of an angle is the number of degrees that one ray of the angle has rotated about the vertex. The red arc shows which angle is being measured.

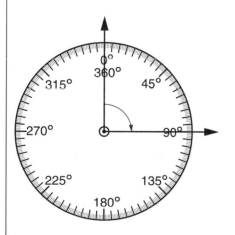

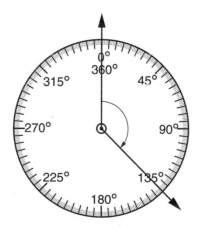

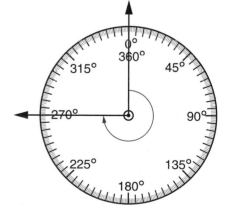

This angle measures 90°. One ray of the angle has rotated 90 degrees, a $\frac{1}{4}$ turn. 90 units, each measuring 1°, fill the space inside the two rays of the angle. A 90° angle is called a **right angle.**

This angle measures 135°. One ray of the angle has rotated 135 degrees. 135 units, each measuring 1°, fill the space inside the two rays of the angle.

This angle measures 270°. One ray of the angle has rotated 270 degrees, or $\frac{3}{4}$ turn of a full circle. 270 units, each measuring 1°, fill the space inside the two rays of the angle.

Measuring and Drawing Angles

Angles are measured with a tool called a **protractor.** You will find both a full-circle and a half-circle protractor on your Geometry Template. Since there are 360 degrees in a circle, a 1° angle marks off $\frac{1}{360}$ of a circle.

The full-circle protractor on the Geometry Template is marked off in 5° intervals from 0° to 360°. It can be used to measure angles, but it *cannot* be used to draw angles. Sometimes you will use a full-circle protractor that is a paper cut-out. This *can* be used to draw angles.

The half-circle protractor on the Geometry Template is marked off in 1° intervals from 0° to 180°.

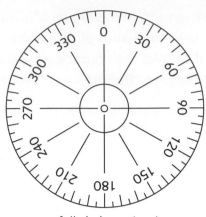

full-circle protractor

It has two scales, each of which starts at 0°. One scale is read clockwise, and the other is read counterclockwise. Before you use a protractor to measure an angle, estimate the size of the angle. Is it **acute** (the measure is less than 90°), close to being a **right angle** (the measure is about 90°), or **obtuse** (the measure is more than 90° but less than 180°)? Use these estimates to make sure your measurement from the protractor makes sense. If the angle is acute, read the smaller number on the half-circle protractor. If it is obtuse, read the larger number.

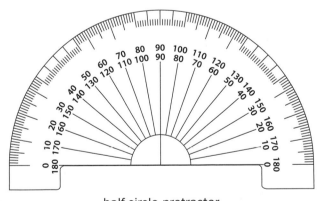

half-circle protractor

Two rays starting from the same endpoint form two angles. The smaller angle, the non-reflex angle, measures between 0° and 180°. The larger angle is called a **reflex angle,** and measures between 180° and 360°. The sum of the measures of the non-reflex angle and the reflex angle is 360°, as they always create a full circle. The red arc tells you the angle you will measure: reflex or non-reflex. If there is no arc, you can assume that you should measure the non-reflex angle.

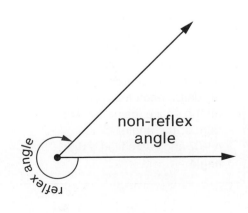

Measuring an Angle with a Full-Circle Protractor

Example

Use the full-circle protractor to measure angle A.

Step 1 Place the hole in the center of the protractor over the vertex of the angle, point A.

Step 2 Line up the 0° mark with one side of the angle so that you can measure the angle clockwise. Make sure that the hole stays over the vertex.

Step 3 Read the degree measure at the mark on the protractor that lines up with the second side of the angle. This is the measure of the angle. The measure of ∠A is 45°.

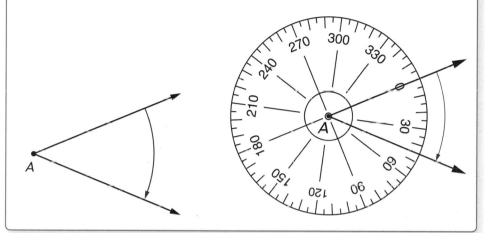

Did You Know?

The way we measure angles was invented by the Babylonians, who lived about 3,000 years ago in what is now the country of Iraq. They counted a year as having 360 days. They used this same number, 360, for measuring angles in a circle because it was an easy multiple in their base-60 system (6 * 60 = 360).

Check Your Understanding

Use your full-circle protractor to measure angles B and C to the nearest degree.

1.

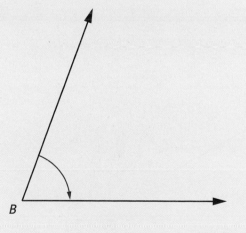

2.

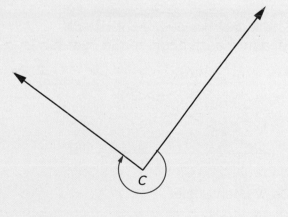

Check your answers in the Answer Key.

Measuring an Angle with a Half-Circle Protractor

To measure angle *PQR* with a half-circle protractor:

Step 1 Lay the baseline of the protractor on $\overrightarrow{QR}$

Step 2 Slide the protractor so that the center of the baseline is over point *Q*, the vertex of the angle.

Step 3 Read the degree measure where $\overrightarrow{QP}$ crosses the edge of the protractor. There are two scales on the protractor. Use the scale that makes sense for the size of the angle you are measuring. The measure of ∠*PQR* is 50°.

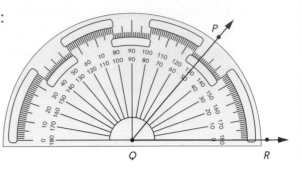

Drawing an Angle with a Half-Circle Protractor

To draw a 40° angle:

Step 1 Draw point *A* and a ray from point *A*.

Step 2 Lay the baseline of the protractor on the ray.

Step 3 Slide the protractor so that the center of the baseline is over point *A*.

Step 4 Make a mark at 40° near the protractor. There are two scales on the protractor. Use the scale that makes sense for the size of the angle you are drawing.

Step 5 Draw a ray from point A through the mark.

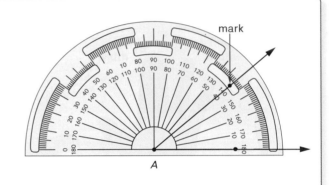

mark

Check Your Understanding

Measure each angle to the nearest degree.

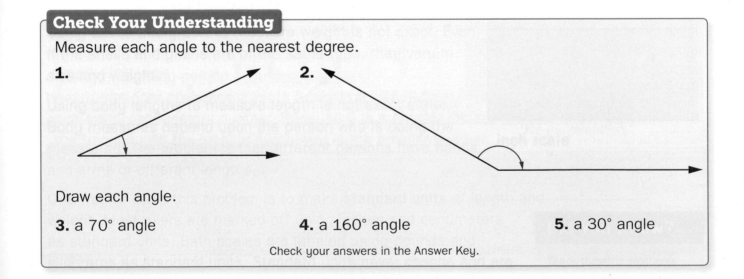

1.

2.

Draw each angle.

3. a 70° angle   **4.** a 160° angle   **5.** a 30° angle

Check your answers in the Answer Key.

Finding Unknown Angle Measures

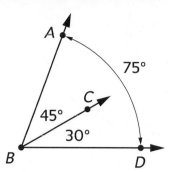

When an angle is decomposed, or broken into smaller angles, the angle measure of the largest angle is equal to the sum of the smaller non-overlapping angle measures.

∠ABD is the largest angle in this diagram.

measure of ∠ABD = measure of ∠ABC + measure of ∠CBD

$$75° \quad = \quad 45° \quad + \quad 30°$$

Sometimes you can find unknown angle measures without a protractor. You can use information in diagrams of angles, angle attributes, measures of benchmark angles, and addition and subtraction to find unknown angle measures.

- Identify the unknown angle measure in the diagram.

- To find the unknown angle measure, create an addition or subtraction equation.

Example

Find the measure of the unknown angle in the diagram using equations.

∠EFG is the largest angle. It is a right angle, so the sum of the smaller angle measures is 90 degrees.

$$50° + d = 90° \qquad\qquad 90° - 50° = d$$
$$50° + \mathbf{40°} = 90° \quad \text{OR} \quad 90° - 50° = \mathbf{40°}$$
$$d = \mathbf{40°} \qquad\qquad d = \mathbf{40°}$$

The measure of ∠HFG = **40°**.

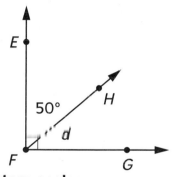

When the sum of two angles is 90°, they are called **complementary angles.** The two smaller angles in this example are complementary angles.

Example

Find the measure of the unknown angle in the diagram using equations.

∠JKL is the largest angle. It is a straight angle, so the sum of the smaller angle measures is 180°.

$$70° + p = 180° \qquad\qquad 180° - 70° = p$$
$$70° + \mathbf{110°} = 180° \quad \text{OR} \quad 180° - 70° = \mathbf{110°}$$
$$p = \mathbf{110°} \qquad\qquad p = \mathbf{110°}$$

The measure of ∠MKL = **110°**.

When the sum of two angles is 180°, they are called **supplementary angles.** The two smaller angles in this example are supplementary angles.

Example

Find the measure of the unknown angle in the diagram using equations.

∠NPQ is the largest angle. It measures 82°, so the sum of the smaller angle measures is 82°.

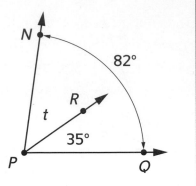

$t + 35° = 82°$ $82° - 35° = t$

47° + 35° = 82°   OR $82° - 35° =$ **47°**

$t =$ **47°** $t =$ **47°**

         The measure of ∠NPR = **47°**.

Check Your Understanding

Find the measures of the unknown angles in each diagram below using equations.

1.

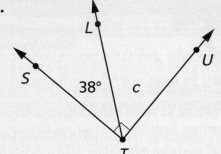

2.

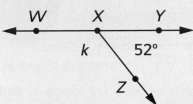

3.

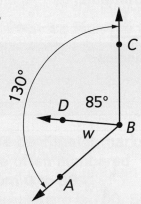

Check your answers in the Answer Key.

Tally Charts and Bar Graphs

There are different ways that you can collect information. You can count, measure, ask questions, or observe something and describe what you see. The information that you collect is called **data.**

You can use a **tally chart** to record and organize data.

> ### Example
>
> Mr. Davis asked each student to name his or her favorite drink. He recorded the students' choices in the tally chart on the right.
>
> There are 25 tally marks in the chart.
>
> That means that 25 different students voted for a favorite drink.
>
> **Favorite Drinks**
>
> | Drink | Tallies |
> |---|---|
> | Milk | 卌 |
> | Chocolate milk | /// |
> | Soft drink | 卌 卌 / |
> | Apple juice | /// |
> | Tomato juice | / |
> | Water | // |

A **bar graph** is a drawing that uses bars to represent data.

> ### Example
>
> This bar graph shows the same information as the tally chart above, but in a different way. This is a scaled bar graph. The scale shows intervals of 2.
>
> The title shows the subject of the graph.
>
>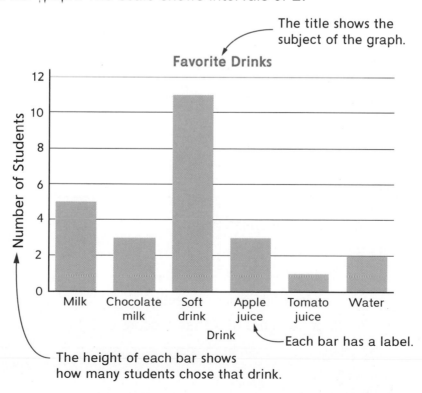
>
> Each bar has a label.
>
> The height of each bar shows how many students chose that drink.

Line Plots

A **line plot** uses check marks, Xs, or other marks to show counts. There is one mark for each count.

Example

The students in Mr. Jackson's class got the scores below on a five-word spelling test. Each score is the number of words a child spelled correctly.

5   3   5   0   4   4   5   4   4   4   2   3   4   5   3   5   4   3   4   4

They drew a line plot to show the data.

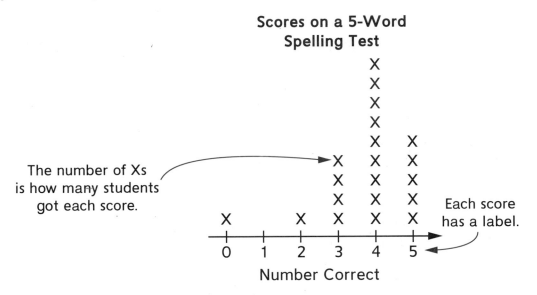

There are 4 Xs above 3. Four children got a score of 3 on the test.

You can get a lot of information from this line plot.

• The most common score on the test was 4.

• All but 2 children spelled 3 or more words correctly.

Check Your Understanding

Use the line plot above to answer the questions.

1. How many students took the spelling test?

2. How many students spelled all five words correctly?

3. How many students spelled just one word correctly?

4. How many more students got a score of 4 than a score of 3?

Scaled Line Plots

You can figure out the scale of a line plot by looking at a set of data. Identify the smallest (minimum) and largest (maximum) numbers in your data set. These can be the first and last numbers on your scale. Then figure out a way to represent the rest of your data. The numbers on your scale should have the same interval, or space, between them.

Example

Scientists measured stag beetles in Europe. Here are their measurements (in inches) in order from smallest to largest:

$1\frac{1}{2}$ $1\frac{1}{2}$ $1\frac{1}{2}$ 2 2 2 2 2 2 2 2

2 $2\frac{1}{2}$ $2\frac{1}{2}$ $2\frac{1}{2}$ $2\frac{1}{2}$ $2\frac{1}{2}$ $2\frac{1}{2}$ $2\frac{1}{2}$ 3 3 $3\frac{1}{2}$

The line plot below shows the data.

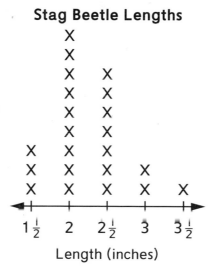

Stag Beetle Lengths

Length (inches)

The line plot shows measurement data scaled in $\frac{1}{2}$-inch intervals.

Check Your Understanding

1. Chad planted sunflower seeds. Two weeks later, he measured the heights of the plants. Here are his results:

   $4\frac{1}{2}$ in. $3\frac{3}{4}$ in.   4 in. $3\frac{3}{4}$ in. $3\frac{3}{4}$ in.

   $3\frac{1}{2}$ in.   4 in. $4\frac{1}{2}$ in.   4 in. $3\frac{3}{4}$ in.

   a. Put the measurements in order from smallest to largest. What is the smallest value and largest value in the data set?

   b. Draw a line plot for the data. Which numbers did you use as the first and last numbers on your scale?

   c. What is the size of the interval (or space) between the numbers on your scale?

   d. Do you have any measurements that do not have Xs? Which one(s)?

   Check your answers in the Answer Key.

Measuring Mass and Volume

Two other quantities that scientists often need to measure are mass and volume. **Mass,** the amount of matter in an object, can be measured with a set of standard masses and a pan balance. Pan balances can measure the mass of an object to the nearest gram, kilogram, or other unit of mass.

Scientists often need to measure the mass of very small amounts of chemicals to make solutions for experiments. These measurements require instruments that are more precise. This analytical scale can measure amounts to the nearest tenth of a milligram or $\frac{1}{10,000}$ of a gram.

Volume, the amount of space occupied by an object, can be measured in a variety of ways. When scientists need to use a precise amount of liquid in an experiment, they can use a graduated cylinder like this one.

Some scientists, like geologists who investigate the natural world, often measure the volume of odd-shaped objects such as rocks. The volume of a rock can be found using a graduated cylinder and water, as shown at the right. Start with a known volume of water, such as 200 milliliters, and then carefully add the rock. After adding the rock, the graduated cylinder reads 235 milliliters. The change in volume when the rock is added is 35 milliliters, which is also the volume of the rock.

(t) ©Iconotec/Alamy; (c) McGraw-Hill Education/Digital Light Source: Richard Hutchings; (tcr) The McGraw-Hill Companies, Inc./Stephen Frisch, photographer

Saving the Greenback

The greenback cutthroat trout is the state fish of Colorado. It was thought to be extinct until it was observed in Rocky Mountain National Park in 1957. Because the greenback cutthroat is now on the endangered species list, environmental scientists are making efforts to ensure its survival by maintaining its habitat.

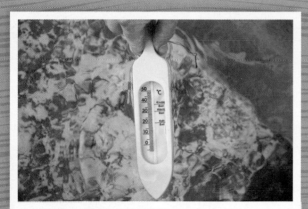

In order to maintain a healthy habitat, scientists monitor stream quality. One important indicator of trout survival is water temperature. The greenback is a cold-water species that cannot survive for long in streams where the water temperature is above 68°F (20°C). The optimal temperature for their growth and sustainability is between 56°F and 57°F.

Just like humans, fish need oxygen to survive. Fish breathe oxygen that is dissolved in stream water. Greenbacks need at least 6 milligrams of oxygen per liter of water in order to thrive. Scientists can monitor the level of dissolved oxygen with instruments like this one.

Another indicator of a stream's quality is the abundance of aquatic insects like the stonefly. Insects like these are the primary food source for the greenback. Scientists collect, count, and measure these insects to monitor their size and availability.

This stonefly nymph measures 1.7 cm.

Weather Measurements

There are many ways to measure what is happening with the weather. We can use a thermometer to measure the temperature or a rain gauge to measure precipitation.

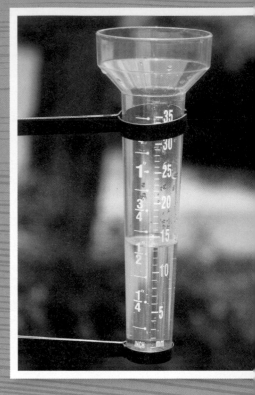

With instruments like this one, scientists can measure wind speed and direction.

Another way to investigate the weather is to measure what is happening in the atmosphere, or the air surrounding Earth. Humidity, the amount of water present in the atmosphere, is important because it has a big effect on how the weather feels. High humidity makes us feel hotter in the summer and colder in the winter. Humidity can be measured with a hygrometer like the one shown at the right.

Atmospheric pressure is the force per unit area exerted on a surface by the mass of the air above it. This force can be measured with a barometer and is sometimes known as barometric pressure. Barometric pressure is measured in millibars (mb). Normal barometric pressure at sea level is about 1,000 millibars. What do you think happens to barometric pressure as you climb a mountain?

When the barometric pressure changes, you probably won't feel it, but you may notice a change in the weather. High barometric pressure above 1,020 millibars usually brings fair weather and clear skies, while low pressure of 970 millibars may mean a storm is on the way.

high pressure

low pressure

Tornadoes are small but violent storms in which winds spin around a system of low atmospheric pressure. The average tornado is about $\frac{1}{2}$ mile wide and lasts no more than an hour. The strongest tornadoes can cause great destruction with wind speeds of more than 300 miles per hour (mph).

Hurricanes are another type of storm in which strong winds rotate around a region of low atmospheric pressure. However, hurricanes are much larger, ranging in size from 100 miles to 1,000 miles wide. While the strongest hurricanes can produce wind speeds of more than 200 mph, wind speeds in hurricanes are usually less than 180 mph. Much of a hurricane's devastation is caused by heavy rains and flooding.

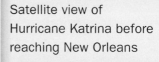

Satellite view of Hurricane Katrina before reaching New Orleans

New Orleans after Hurricane Katrina

Scientific Measurements and Your Health

Scientific measurements are made every day to monitor the quality of the air we breathe and the water we drink. Clean air is essential to good health. Air quality monitoring stations like the one to the right can produce reliable hourly measurements of pollutants including ozone, nitrogen oxides, sulfur dioxide, carbon monoxide, and particulates.

Clean drinking water is also necessary for good health. Most people in the United States receive water from a community water system that provides its customers with an annual water quality report. These reports contain information about the kinds and amounts of contaminants found in the system's drinking water. These measurements are made at water treatment or testing facilities.

What questions do you have about the natural world? What measurements could you make to answer your questions?

You can also take measures of personal heath. Two common measures of health are body temperature and pulse, or how fast your heart is beating. Normal body temperature is 98.6°F and the average pulse rate for people 10 years and older is between 60 and 100 beats per minute (bpm). You can use a thermometer to measure you temperature. To measure your pulse, find a place on your wrist or neck where you can feel your pulse. Use a clock with a second hand to count the number of beats in 15 seconds. Multiply that number by 4 (because 4 * 15 seconds = 60 seconds, or 1 minute) to find your heart rate in beats per minute.

Geometry

Geometry in Our World

The world is filled with geometry. There are angles, segments, lines, and curves everywhere you look. There are 2-dimensional and 3-dimensional shapes of every type.

Many wonderful geometric patterns can be seen in nature. You can find patterns in flowers, spider webs, leaves, seashells, even your own face and body.

The ideas of geometry are also found in the things people create. Think of the games you play. Checkers is played with round pieces. The game board is covered with squares. Basketball and tennis are played with spheres. They are played on rectangular courts that are painted with straight and curved lines. The next time you play or watch a game, notice how geometry is important to the way the game is played.

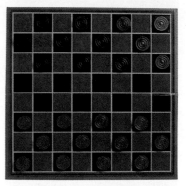

The places we live in are built from plans that use geometry. Buildings almost always have rectangular rooms. Outside walls and roofs often include sections that have triangular shapes. Archways are curved and are often shaped like semicircles (half circles). Staircases may be straight or spiral. Buildings and rooms are often decorated with beautiful patterns. You see these decorations on doors and windows; on walls, floors, and ceilings; and on railings of staircases.

The clothes people wear are often decorated with geometric shapes. So are the things they use every day. Everywhere in the world, people create things using geometric patterns. Examples include quilts, tiles, baskets, and pottery. Some patterns are shown here. Which are your favorites?

Make a practice of noticing geometric shapes around you. Pay attention to bridges, buildings, and other structures. Look at the ways in which simple shapes such as triangles, rectangles, and circles are combined. Notice interesting designs. Share these with your classmates and your teacher.

In this section, you will learn about lines, angles, geometric shapes, and symmetry. As you learn, try to find examples of geometric figures around you.

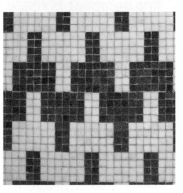

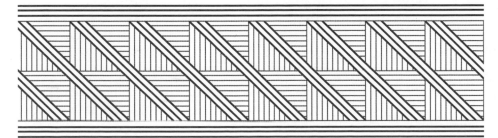

Points and Line Segments

A **point** is a location in space. You often make a dot with a pencil to show where a point is.

Letters are used to name points. The letter names make it easy to talk about the points. For example, in the illustration at the right, point *A* is closer to point *B* than it is to point *P*. Point *P* is closer to point *B* than it is to point *A*.

A **line segment** is made up of two points and the straight path between them. You can use any tool with a straight edge to draw the path between two points.

- The two points are called the **endpoints** of the line segment.
- The line segment follows the shortest path between the endpoints.

The line segment below is called *line segment AB* or *line segment BA.*

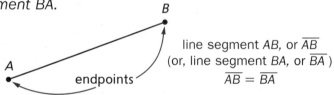

line segment *AB*, or $\overline{AB}$
(or, line segment *BA*, or $\overline{BA}$)
$\overline{AB} = \overline{BA}$

The symbol for a line segment is a raised bar. The bar is written above the letters that name the endpoints for the segment. The name of the line segment above can be written $\overline{AB}$ or $\overline{BA}$.

Straightedge and Ruler

A **straightedge** is a strip of wood, plastic, or metal that may be used to draw a straight path. A **ruler** is a straightedge that is marked so that it may be used to measure lengths.

- Every ruler is a straightedge.
- However, every straightedge is not a ruler.

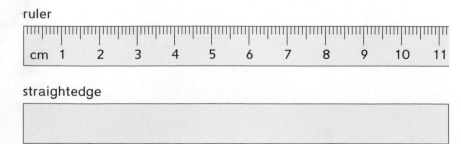

Rays and Lines

A **ray** is a straight path that starts at one endpoint and goes on forever in *one* direction.

To draw a ray, draw a line segment and extend the path beyond one endpoint. Then add an arrowhead to show that the path goes on forever.

The ray at the right is called *ray RA*.

Point *R* is the endpoint of ray *RA*. The endpoint is always the first letter in the name of a ray. The second letter can be any other point on the ray.

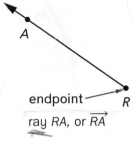

endpoint

R

ray *RA*, or $\overrightarrow{RA}$

The symbol for a ray is a raised arrow, pointing to the right. For example, ray *RA* can be written $\overrightarrow{RA}$.

A **line** is a straight path that goes on forever in *both* directions.

To draw a line, draw a line segment and extend the path beyond each endpoint. Then add an arrowhead at each end.

The symbol for a line is a raised bar with two arrowheads. The name of the line at the right can be written either as $\overleftrightarrow{EF}$ or as $\overleftrightarrow{FE}$. You can name a line by listing any two points on the line, in any order.

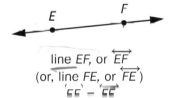

line *EF*, or $\overleftrightarrow{EF}$
(or, line *FE*, or $\overleftrightarrow{FE}$)

Example

Write all the names for this line.

Points *C*, *A*, and *B* are all on the line.

Use any two points to write the name of the line:

$\overleftrightarrow{CA}$ or $\overleftrightarrow{AC}$ or $\overleftrightarrow{CB}$ or $\overleftrightarrow{BC}$ or $\overleftrightarrow{AB}$ or $\overleftrightarrow{BA}$

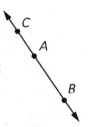

Check Your Understanding

1. Draw and label $\overleftrightarrow{GH}$.
2. Draw and label a point *T* that is not on $\overleftrightarrow{GH}$.
3. Draw and label $\overrightarrow{TG}$ and $\overrightarrow{TH}$.

Check your answers in the Answer Key.

Angles

An **angle** is formed by two rays, two line segments, or a ray and line segment that share the same endpoint.

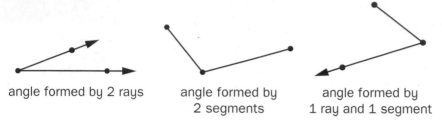

angle formed by 2 rays

angle formed by 2 segments

angle formed by 1 ray and 1 segment

The endpoint where the rays or segments meet is called the **vertex** of the angle. The rays or segments are called the **sides** of the angle.

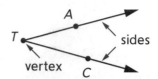

Naming Angles

The symbol for an angle is ∠. An angle can be named in two ways:

1. Name the vertex. The angle shown above is angle *T*. Write this as ∠*T*.

2. Name three points: the vertex and one point on each side of the angle. The angle above can be named angle *ATC* (∠*ATC*) or angle *CTA* (∠*CTA*). The vertex must always be listed in the middle, between the points on the sides.

Angle Measures

The size of an angle is the amount of turning about the vertex from one side of the angle to the other. Angles are measured in degrees. A **degree** is a unit of measure for the size of an angle. A full turn about a vertex makes an angle that measures 360 degrees. A **protractor** is a tool used to measure angles. There are two types of protractors: full-circle protractors and half-circle protractors.

The degree symbol ° is often used in place of the word *degrees*. The measure of ∠*T* above is 30 degrees or 30°.

The small curved arrow in each picture shows which angle opening to measure.

full-circle protractor

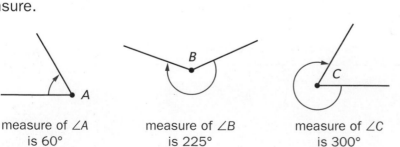

measure of ∠*A* is 60°

measure of ∠*B* is 225°

measure of ∠*C* is 300°

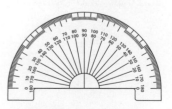

half-circle protractor

Classifying Angles

Angles may be classified according to size.

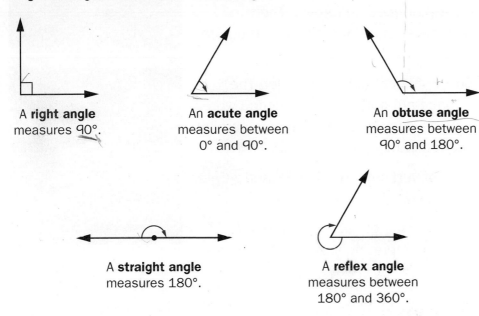

A **right angle**
measures 90°.

An **acute angle**
measures between
0° and 90°.

An **obtuse angle**
measures between
90° and 180°.

A **straight angle**
measures 180°.

A **reflex angle**
measures between
180° and 360°.

A **right angle** is an angle whose sides form a square corner. You may draw a small corner symbol inside an angle to show that it is a right angle.

A **straight angle** is an angle whose sides form one straight path.

Check Your Understanding

1. Draw a right angle.
2. Draw an obtuse angle.
3. Refer to the figure at the right.
   a. Which angles are right angles? ∠A
   b. Which angles are acute angles? ∠E
   c. Which angles are obtuse angles? ∠G
   d. Which angles are reflex angles? ∠D
   e. Which angles are straight angles? ∠F
   f. Give another name for ∠E.

acute angle

Check your answers in the Answer Key.

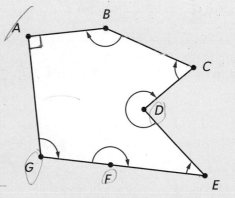

Parallel Lines and Segments

Parallel lines are lines on a flat surface that never cross or meet. Think of a straight railroad track that goes on forever. The two rails are parallel lines. The rails never meet or cross, and they are always the same distance apart.

Parallel line segments are parts of lines that are parallel. The top and bottom edges of this page are parallel. If each edge were extended forever in both directions, the lines would remain parallel because they would always be the same distance apart.

The symbol for parallel is a pair of vertical lines ||. If $\overline{FE}$ and $\overline{JK}$ are parallel, write $\overline{FE} \parallel \overline{JK}$.

If lines or segments cross or meet each other, they **intersect.** Lines or segments that intersect and form right angles are called **perpendicular** lines or segments.

The symbol for perpendicular is ⊥ which looks like an upside-down letter T. If $\overleftrightarrow{RS}$ and $\overleftrightarrow{XY}$ are perpendicular, write $\overleftrightarrow{RS} \perp \overleftrightarrow{XY}$.

Examples

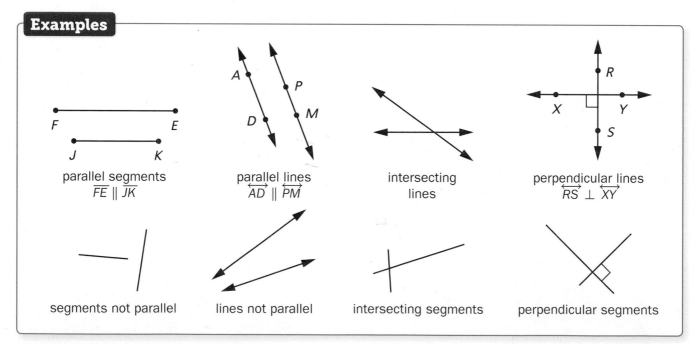

parallel segments
$\overline{FE} \parallel \overline{JK}$

parallel lines
$\overleftrightarrow{AD} \parallel \overleftrightarrow{PM}$

intersecting lines

perpendicular lines
$\overleftrightarrow{RS} \perp \overleftrightarrow{XY}$

segments not parallel

lines not parallel

intersecting segments

perpendicular segments

Check Your Understanding

Draw and label the following.

1. Parallel line segments *EF* and *GH*

2. A line segment that is perpendicular to both $\overline{EF}$ and $\overline{GH}$

Check your answers in the Answer Key.

Line Segments, Rays, Lines, and Angles

Figure	Symbol	Name and Description
•A	A	**point:** a location in space
E ↗ F endpoints	$\overline{EF}$ or $\overline{FE}$	**line segment:** a straight path between two points called its endpoints
N M endpoint → M	$\overrightarrow{MN}$	**ray:** a straight path that goes on forever in one direction from an endpoint
P R	$\overleftrightarrow{PR}$ or $\overleftrightarrow{RP}$	**line:** a straight path that goes on forever in both directions
vertex S T P	$\angle T$ or $\angle STP$ or $\angle PTS$	**angle:** two rays or line segments with a common endpoint, called the vertex
A B S R	$\overleftrightarrow{AB} \parallel \overleftrightarrow{RS}$	**parallel lines:** lines that never cross or meet and are everywhere the same distance apart
	$\overline{AR} \parallel \overline{RS}$	**parallel line segments:** segments that are parts of lines that are parallel
R E D S	none	**intersecting lines:** lines that cross or meet
	none	**intersecting line segments:** segments that cross or meet
R B S C	$\overleftrightarrow{BC} \perp \overleftrightarrow{RS}$	**perpendicular lines:** lines that intersect at right angles
	$\overline{BC} \perp \overline{RS}$	**perpendicular line segments:** segments that intersect at right angles

Check Your Understanding

Draw and label each of the following.

1. point M

2. $\overleftrightarrow{RT}$

3. $\angle TRY$

4. $\overline{XY}$

5. $\overline{DE} \parallel \overline{KL}$

6. $\overrightarrow{FG}$

Check your answers in the Answer Key.

Polygons

A **polygon** is a flat, 2-dimensional figure made up of line segments called **sides**. A polygon can have any number of sides, as long as it has at least three sides.

- The sides of a polygon are connected end to end and make one closed path.
- The sides of a polygon do not cross.

Each endpoint where sides meet is called a **vertex**. The plural of the word *vertex* is **vertices**.

Figures That Are Polygons

4 sides, 4 vertices

3 sides, 3 vertices

7 sides, 7 vertices

Figures That Are NOT Polygons

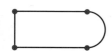
All sides of a polygon must be line segments. Curved lines are not line segments.

The sides of a polygon must form a closed path.

A polygon must have at least 3 sides.

The sides of a polygon must not cross.

Polygons are named based on their number of sides. The prefix for a polygon's name tells the number of sides it has.

Prefixes

tri- 3 quad- 4 penta- 5 hexa- 6 hepta- 7

triangle
3 sides

quadrilateral
4 sides

pentagon
5 sides

hexagon
6 sides

heptagon
7 sides

octa- 8 nona- 9 deca- 10 dodeca- 12

octagon
8 sides

nonagon
9 sides

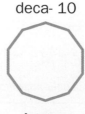
decagon
10 sides

dodecagon
12 sides

Triangles

Triangles have fewer sides and angles than any other polygon. The prefix *tri-* means *three*. All triangles have three vertices, three sides, and three angles.

For the triangle shown here:

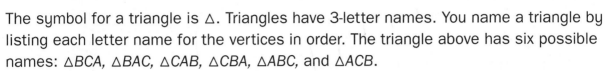

Angle *A* is formed by sides that meet at vertex *A*.

- The vertices are the points *B*, *C*, and *A*.
- The sides are $\overline{BC}$, $\overline{BA}$, and $\overline{CA}$.
- The angles are $\angle B$, $\angle C$, and $\angle A$.

The symbol for a triangle is △. Triangles have 3-letter names. You name a triangle by listing each letter name for the vertices in order. The triangle above has six possible names: △*BCA*, △*BAC*, △*CAB*, △*CBA*, △*ABC*, and △*ACB*.

Triangles have many different sizes and shapes. You will work with two types of triangles that have special names.

A **right triangle** is a triangle that contains one right angle (square corner). In a right triangle, the sides that form the right angle are perpendicular to each other. Triangle *RDP* is a right triangle because line segments *RD* and *RP* are perpendicular to each other.

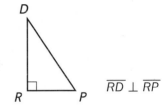

$\overline{RD} \perp \overline{RP}$

Right triangles have many different shapes and sizes.

An **equilateral triangle** is a triangle whose three sides all have the same length. Equilateral triangles have many different sizes, but all equilateral triangles have the same shape.

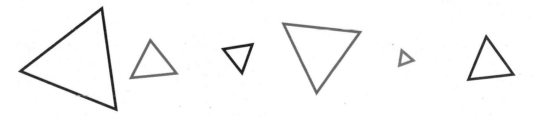

A right triangle cannot be an equilateral triangle because the side opposite the right angle is always longer than each of the other sides.

Quadrilaterals

A **quadrilateral** is a polygon with four sides. The prefix *quad-* means *four*. All quadrilaterals have four vertices, four sides, and four angles. Another name for quadrilateral is **quadrangle.**

For the quadrilateral shown here:

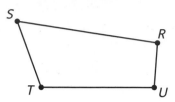

- The sides are $\overline{RS}$, $\overline{ST}$, $\overline{TU}$, and $\overline{UR}$.
- The vertices are R, S, T, and U.
- The angles are ∠R, ∠S, ∠T, and ∠U.

A quadrilateral is named by listing in order the letter names for the vertices. The quadrilateral above has eight possible names:

RSTU, RUTS, STUR, SRUT, TURS, TSRU, URST, UTSR

Some quadrilaterals have *at least* one pair of parallel sides. These quadrilaterals are called **trapezoids.**

Reminder: Two sides are parallel if they never meet, no matter how far they are extended in either direction.

Figures That Are Trapezoids

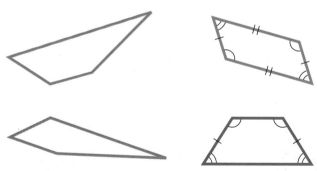

Each figure has *at least* one pair of parallel sides.

Figures That Are NOT Trapezoids

no parallel sides   not a quadrilateral   no parallel sides

Side or Angle Markings

When the markings on the sides of a polygon are the same, it means the sides are the same length. When the markings on angles of a polygon are the same, it means that the angles have the same measure.

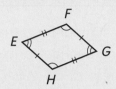

In quadrilateral *EFGH*, side $\overline{EF}$ has the same length as side $\overline{GH}$, and side $\overline{HE}$ has the same length as side $\overline{FG}$. ∠E has the same angle measure as ∠G, and ∠F has the same angle measure as ∠H.

Did You Know?

In mathematics, there are times when concepts or ideas are defined differently. In *Everyday Mathematics*, a **trapezoid** is defined as a quadrilateral that has *at least* one pair of parallel sides. With this definition, parallelograms are also trapezoids. Other sources define a trapezoid as a quadrilateral with *exactly* one pair of parallel sides. With this definition, parallelograms are not trapezoids since parallelograms have more than one pair of parallel sides.

Quadrilaterals: Examples and Definitions

Many special types of quadrilaterals have been given names.

Quadrilateral Type	Example	Definition
trapezoid		**Trapezoids** are quadrilaterals with at least one pair of parallel sides.
isosceles trapezoid		**Isosceles trapezoids** are trapezoids with a pair of consecutive angles with the same measure. Consecutive angles are two angles that have one side in common.
kite		**Kites** are quadrilaterals whose four sides are in two pairs of equal length, with the equal-length sides next to each other. If all four sides of a kite are the same length, the kite is also a rhombus.
parallelogram		**Parallelograms** are quadrilaterals that have two pairs of parallel sides.
rectangle		**Rectangles** are parallelograms that have four right angles. The sides do not all have to be the same length.
rhombus		**Rhombuses** are parallelograms with four sides that have the same length.
square		**Squares** are parallelograms with four right angles and four sides that have the same length.

Check Your Understanding

1. What is the difference between a square and a rectangle?
2. What types of quadrilaterals are both a trapezoid and a kite? How do you know?

Check your answers in the Answer Key.

The Geometry Template

The **Geometry Template** has many uses.

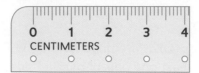

The template has two rulers. The inch scale measures in inches and fractions of an inch. The centimeter scale measures in centimeters and millimeters. Use either side of the template as a straightedge for drawing line segments.

There are 17 different geometric figures on the template. The figures labeled "PB" are **pattern-block shapes.** These are half the size of real pattern blocks. There is a hexagon, a trapezoid, two different rhombuses, an equilateral triangle, and a square. These will come in handy for some of the activities you will do this year.

Each triangle on the template is labeled with the letter T and a number. Triangle "T1" is an equilateral triangle whose sides all have the same length. Triangles "T2" and "T5" are right triangles. Triangle "T3" has sides that all have different lengths. Triangle "T4" has two sides of the same length.

The remaining shapes are circles, squares, a regular octagon, a regular pentagon, a kite, a rectangle, a parallelogram, and an ellipse.

The two circles near the inch scale can be used as ring-binder holes. Use these to store your template in your notebook.

Use the **half-circle** and **full-circle protractors** at the bottom of the template to measure and draw angles.

> ## Did You Know?
>
> Early in the 17th century, the German astronomer and mathematician Johannes Kepler showed that the orbit of a planet about the sun is an ellipse.

Check Your Understanding

Use your Geometry Template to trace each of the shapes.
Then try to label all the shapes with their names without looking at page 237.

Check your answers on page 237.

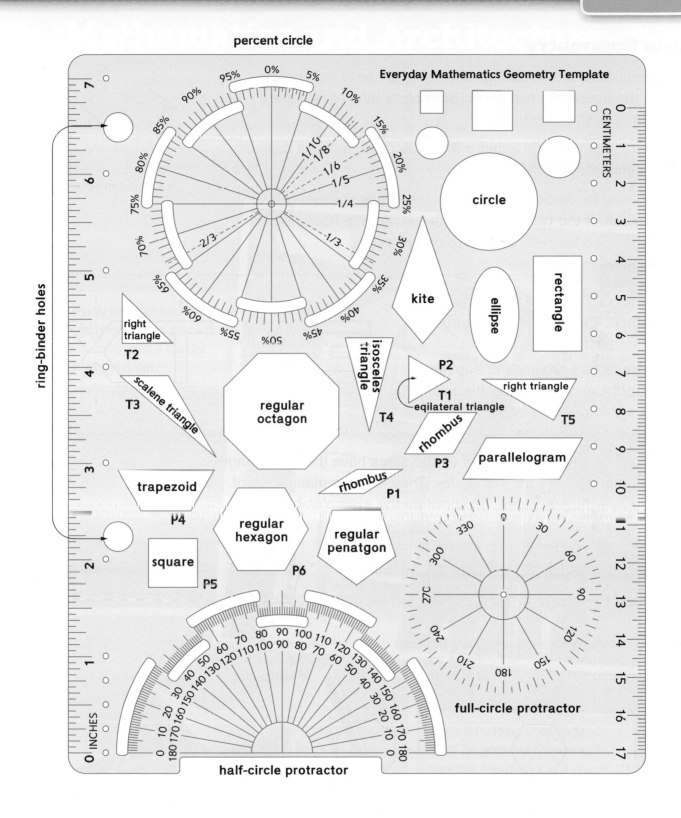

percent circle

Everyday Mathematics Geometry Template

CENTIMETERS

ring-binder holes

95% 0% 5%
90% 10%
85% 15%
80% 1/10 20%
75% 1/8
1/6
1/5 25%
70% 1/4
2/3 1/3 30%
65% 35%
60% 40%
55% 45%
50%

circle

kite

ellipse

rectangle

right triangle
T2

scalene triangle
T3

isosceles triangle

regular octagon

P2
T1
eqilateral triangle

right triangle
T5

rhombus
P3

parallelogram

T4

rhombus
P1

trapezoid
P4

regular hexagon
P6

regular penatgon

square
P5

full-circle protractor

330 0 30
300 60
270 90
240 120
210 150
180

half-circle protractor

INCHES

Homes

People everywhere need places to live, but their homes can be quite different. The shape and structure of homes are influenced by the materials available and the way people live.

Some Mayans built round or rectangular huts called *chozas* or *nah*. The walls were first built with vertical wood sticks then covered with mud or clay.

Native Americans of New Mexico used mud bricks to build villages of connected homes. These homes are in Taos, New Mexico. Why do you think the rooms are shaped like cubes?

Pieces of straw mixed with mud make the bricks durable.

The roof inside of this rectangular choza is made with a grid of wooden poles covered with layers of grass to keep the home dry.

This home in France has geometric patterns on the walls. The steep angle of the straw roof lets rain and snow run off easily.

Homes for Powerful People

Homes for a country's leaders or royalty are designed to be awe-inspiring and beautiful. Many important buildings use symmetry to create a feeling of balance and order.

Bodiam castle in Sussex, England was built in 1385 for a royal councilor to the King of England. Notice the symmetry to the left and right sides of the bridge.

The Grand Palace in Bangkok, Thailand was built in 1782. Note the many different geometrical shapes and patterns in this complex structure.

The White House is home to the president of the United States. Notice the number of columns and windows on either side of the building.

Ceremonial Buildings

In many societies, the most fantastic architecture can be found in places of worship or monuments for tombs.

In ancient Egypt's old kingdom, the Pharaohs chose the pyramid as the shape for their tombs.

The Taj Mahal is a tomb that was built in India in the 17th century.

The shapes and symmetry of the Heian Jingu Shrine are a common feature of Japanese architecture.

The Malacca Straits Mosque is built on stilts. It blends Middle Eastern features, like the dome, with Malaysian features shown in the side turrets.

Modern Buildings

Frank Gehry and I.M. Pei are two of the many architects who have brought new ideas to architecture in the past century.

The Weisman Art Museum in Minneapolis, Minnesota is another of Gehry's famous works. What geometric shapes can you find in his design?

Frank Gehry, a Canadian-American architect, has designed buildings all over the world. One of his well-known projects is the Walt Disney Concert Hall in Los Angeles, which was completed in 2003.

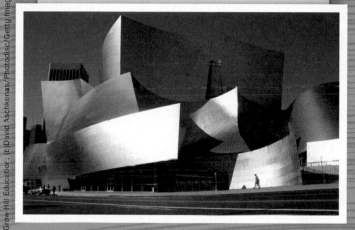

I.M. Pei was born in China, but later emigrated to the United States. He designed the Louvre Pyramid, a glass and metal structure that is the main entrance to the largest art museum in Paris, France.

Pei often combines simple geometric forms in his designs. This is the Rock and Roll Hall of Fame and Museum in Cleveland, Ohio.

Other Structures

How many different shapes can you find in these structures?
Which shapes do you think give strength to the structures?
Which ones do you think add beauty? Which ones do you think
do both? Neither?

Tianjin Eye, Tianjin, China

Gate of Europe towers and Monument
to Calvo Sotelo, Madrid, Spain

Cloud Gate, Chicago, Illinois

Rose Rotana Tower, Dubai,
United Arab Emirates

Games

Throughout the year, you will play games that help you practice important math skills. Playing mathematics games gives you a chance to practice math skills in a different way. We hope that you will play often and have fun.

In this section of your *Student Reference Book*, you will find the directions for many games. The numbers in most games are generated randomly. This means that when you play games over and over, it is unlikely that you will repeat the same problems.

Many students have created their own variations of these games to make them more interesting. We encourage you to do this too.

Materials

The materials for each game are different and may include cards, dice, coins, counters, and calculators. Some games use the set of fraction cards found in the back of your *Math Journal 1* or in your eToolkit. For some games, you will have to make a gameboard or a score sheet. These instructions are included with the game directions. For other games, your teacher will provide gameboards and card decks.

Number Cards. You need a deck of number cards for many of the games. You can use an Everything Math Deck, a deck of regular playing cards, or make your own deck out of index cards. An Everything Math Deck is part of the eToolkit.

An Everything Math Deck includes 54 cards. There are 4 cards each for the numbers 0–10. And there is 1 card for each of the numbers 11–20.

You can also use a deck of regular playing cards after making a few changes. A deck of playing cards includes 54 cards (52 regular cards, plus 2 jokers). To create a deck of number cards, use a permanent marker to mark the cards in the following ways:

- Mark each of the 4 aces with the number 1.
- Mark each of the 4 queens with the number 0.
- Mark the 4 jacks and 4 kings with the numbers 11 through 18.
- Mark the 2 jokers with the numbers 19 and 20.

Fraction Cards

Number Cards

eToolkit

Comstock Images/Alamy

Game	Skill	Page
Angle Add-Up	Drawing angles; finding the measures of unknown angles	248
Angle Race	Recognizing angle measures	249
Angle Tangle	Estimating and measuring angle size	250
Beat the Calculator (Extended Facts)	Practicing extended multiplication facts	251
Buzz Games	Finding multiples of a number and common multiples of two numbers	252
Decimal Top-It	Understanding place value for decimals	253
Divide and Conquer	Practicing extended division facts	254
Division Arrays	Modeling division with and without remainders	255
Division Dash	Dividing 2-digit numbers by 1-digit numbers	256
Factor Bingo	Identifying factors of a number	257
Factor Captor	Finding factors of a number	258
Fishing for Digits	Understanding place value	259
Fishing for Fractions	Adding fractions with like denominators	260
Fraction/Decimal Concentration	Recognizing fractions and decimals that are equivalent	262
Fraction Match	Recognizing equivalent fractions	263
Fraction Multiplication Top-It	Multiplying a whole number by a fraction	264
Fraction Top-It	Comparing fractions	265
How Much More?	Solving comparison number stories	266
Multiplication Wrestling	Multiplying 2-digit numbers using partial-products multiplication	267
Name That Number	Finding equivalent names for numbers using operations	268
Number Top-It	Understanding place value for whole numbers	269
Polygon Capture	Identifying properties of polygons	270
Product Pile-Up	Practicing multiplication facts 1 to 10	271
Rugs and Fences	Finding the area and perimeter of rectangles by applying formulas	272
Spin-and-Round	Rounding 6-digit numbers to the nearest hundred through hundred-thousand	273
Subtraction Target Practice	Subtracting 2-digit numbers	274
Top-It Games	Adding, subtracting, multiplying, and dividing	275

Angle Add-Up

Materials	☐ number cards 1–8 (4 of each)
	☐ number cards 0 and 9 (1 of each)
	☐ 1 straightedge
	☐ 1 *Angle Add-Up* Record Sheet for each player (*Math Masters*, p. G47)
Players	2
Skill	Drawing angles; finding the measures of unknown angles
Object of the Game	To score more points in 3 rounds.

Directions

1 Shuffle the cards and place the deck number-side down on the table.

2 Each round, each player draws the number of cards indicated on the record sheet.

3 Each player uses the number cards to fill in the blanks and form angle measures so the unknown angle measure is as large as possible.

4 Players add or subtract to find the measure of the unknown angle and record it in the circle on their record sheet. The measure of the unknown angle is the player's score for the round.

5 Each player uses a straightedge, pencil, and the full-circle protractor on the record sheet to show that the angle measure of the whole is the sum of the angle measures of the parts.

6 Play 3 rounds for a game. The player with the greater total number of points at the end of 3 rounds wins the game.

Example

In Round 1, Suma draws a 2, 7, 1, and 5. She creates the angle measures 51° and 72°, then finds the sum: m∠ABD + m∠DBC = m∠ABC. She shows the sum 51° + 72° = 123° on her protractor and scores 123 points for the round.

Round 1:
Draw 4 cards. $\underline{\ \ 5\ \ }\ \underline{\ \ 1\ \ }° + \underline{\ \ 7\ \ }\ \underline{\ \ 2\ \ }° = \underline{123}°$

m∠ABD   m∠DBC   m∠ABC

Angle Race

Materials ☐ 1 24-pin circular geoboard or 1 sheet of Circular-Geoboard Paper (*Math Masters*, p. TA51)

☐ 15 rubber bands, or a straightedge and a pencil

☐ 1 set of *Angle Race* Degree-Measure Cards (*Math Masters*, p. G46)

Players 2

Skill Recognizing angle measures

Object of the Game To complete an angle exactly on the 360° mark on a circular geoboard.

Directions

1 Shuffle the cards. Place the deck number-side down.

2 If you have a circular geoboard, stretch a rubber band from the center peg to the 0° peg.

If you do *not* have a circular geoboard, use a sheet of Circular-Geoboard Paper. Draw a line segment from the center dot to the 0° dot. Instead of stretching rubber bands, you will draw line segments.

3 Players take turns using the same geoboard or paper.

4 When it is your turn, select the top card from the deck. Make an angle on the geoboard that has the same degree measure as shown on the card. Use the last rubber band placed on the geoboard as one side of your angle. Make the second side of your angle by stretching another rubber band from the center peg to a peg on the circle, going *clockwise*.

5 Rubber bands may not go past the 360° (or 0°) peg. If you must go past the 360° peg, you lose your turn.

6 The first player to complete an angle exactly on the 360° peg wins.

Circular Geoboard

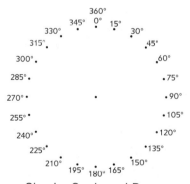

Circular-Geoboard Paper

Example

Nia draws a 30° card. She makes a 30° angle by stretching a rubber band from the center peg to the 30° peg. Max draws a 75° card. He makes a 75° angle by stretching a rubber band from the center peg to the 105° peg, as shown above. They continue to take turns, adding angles clockwise around the circle, until one player reaches the 360° peg exactly.

Angle Tangle

Materials	☐ 1 protractor
	☐ 1 straightedge
	☐ several blank sheets of paper
Players	2
Skill	Estimating and measuring angle size
Object of the Game	To estimate angle sizes accurately and have the lower total score.

Directions

In each round:

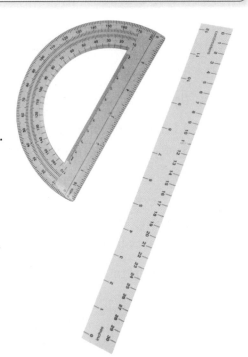

1 Player 1 uses a straightedge to draw an angle on a sheet of paper.

2 Player 2 estimates the degree measure of the angle.

3 Player 1 measures the angle with a protractor. Players agree on the measure.

4 Player 2's score is the difference between the estimate and the actual measure of the angle. (The difference will be 0 or a positive number.)

5 Players trade roles and repeat Steps 1–4.

Players add their scores at the end of five rounds.
The player with the lower total score wins the game.

Example

	Player 1			Player 2		
	Estimate	Actual	Score	Estimate	Actual	Score
Round 1	120°	108°	12	50°	37°	13
Round 2	75°	86°	11	85°	87°	2
Round 3	40°	44°	4	15°	19°	4
Round 4	60°	69°	9	40°	56°	16
Round 5	135°	123°	12	150°	141°	9
Total Score			48			44

Player 2 has the lower total score. Player 2 wins the game.

Beat the Calculator (Extended Facts)

Materials	☐ number cards 1–10 (4 of each)
	☐ 1 calculator
Players	3
Skill	Practicing extended multiplication facts
Object of the Game	To multiply numbers without a calculator faster than a player using one.

Directions

1 One player is the "Caller," one is the "Calculator," and one is the "Brain."

2 Shuffle the deck and place it number-side down on the table.

3 The Caller draws 2 cards from the number deck, attaches a 0 to either one of the factors or to both factors, and asks for the product.

4 The Calculator solves the problem using a calculator. The Brain solves it without a calculator. The Caller decides who got the answer first.

5 The Caller continues to draw 2 cards at a time from the number deck, attach a 0 to one or both factors, and ask for the product.

6 Players trade roles every 10 turns or so.

Example

The Caller turns over a 4 and an 8.

He or she may make up any one of the following problems:

4 * 80   40 * 8   40 * 80

The Brain and the Calculator solve the problem.

The Caller decides who got the answer first.

Buzz Games

Materials	none
Players	5–10
Skill	Finding multiples of a number and common multiples of two numbers
Object of the Game	To correctly say either "BUZZ," "BIZZ," "BIZZ-BUZZ," or the next number when it is your turn.

Buzz

Directions

1 Players sit in a circle and choose a leader. The leader names any whole number from 3 to 9. This number is the BUZZ number. The leader also chooses the STOP number. The STOP number should be at least 30.

2 The player to the left of the leader begins the game by saying "one." Play continues clockwise with each player saying either the next whole number or "BUZZ."

3 A player must say "BUZZ" instead of the next number if:

- The number is the BUZZ number or a multiple of the BUZZ number; or

- The number contains the BUZZ number as one of its digits

4 If a player makes an error, the next player starts with 1.

5 Play continues until the STOP number is reached.

6 For the next round, the player to the right of the leader becomes the new leader.

Example

The BUZZ number is 4. Play should proceed as follows: 1, 2, 3, BUZZ, 5, 6, 7, BUZZ, 9, 10, 11, BUZZ, 13, BUZZ, 15, and so on.

Bizz-Buzz

Directions

Bizz-Buzz is played like *Buzz*, except the leader names 2 numbers: a BUZZ number and a BIZZ number.

Players say:

1 "BUZZ" if the number is a multiple of the BUZZ number.

2 "BIZZ" if the number is a multiple of the BIZZ number.

3 "BIZZ-BUZZ" if the number is a multiple of both the BUZZ number and the BIZZ number.

Example

The BUZZ number is 6, and the BIZZ number is 3. Play should proceed as follows: 1, 2, BIZZ, 4, 5, BIZZ-BUZZ, 7, 8, BIZZ, 10, 11, BIZZ-BUZZ, 13, 14, BIZZ, 16, and so on. The numbers 6 and 12 are replaced by "BIZZ-BUZZ" since 6 and 12 are multiples of both 3 and 6.

Decimal Top-It

Materials	☐ number cards 0–9 (4 of each)
	☐ 1 *Decimal Top-It* Mat (*Math Masters*, p. G30)
	☐ 1 *Top-It* Record Sheet for each player (*Math Masters*, p. G2)
Players	2
Skill	Understanding place value for decimals
Object of the Game	To make the larger 2-digit decimal number.

Directions

1 Shuffle the cards and place the deck number-side down on the table.

2 Each player uses one row of boxes on the *Decimal Top-It* Mat.

3 In each round, players take turns turning over the top card from the deck and placing it on any one of their empty boxes, until each player has taken 2 turns and placed 2 cards on his or her row of the game mat.

4 At the end of each round, players read their numbers aloud and compare them. Each player records the comparison on his or her *Top-It* Record Sheet. The player with the larger number takes all of the cards.

5 The game ends when there are not enough cards left for each player to have another turn. The player with more cards wins.

Example

Kent and Kari played *Decimal Top-It*. Here is the result.

Kari's number is larger than Kent's number. Kari takes all of the cards for this round.

They both record 0.35 < 0.64 on their record sheets.

Decimal Top-It Mat

Ones   Tenths   Hundredths

Kent   0. [3] [5]

Kari   0. [6] [4]

Variation

To play with 3–5 players, use 1 *Decimal Top-It* Mat (*Math Masters*, page G30) for every 2 players. Each player uses one row on a game mat. Players take turns until every player has placed 2 cards on his or her row, then all players read their numbers aloud and compare their numbers. The player with the largest number takes all of the cards. Players do not need to record their comparisons.

Divide and Conquer

Materials	☐ 1 set of *Divide and Conquer* Fact Triangles (*Math Masters*, pp. G38–G40)
	☐ 1 calculator
Players	3
Skill	Practicing extended division facts
Object of the Game	To divide extended division facts correctly without a calculator faster than a player using one.

Directions

1 One player is the "Caller," one is the "Calculator," and one is the "Brain."

2 The Caller shuffles the Fact Triangles and places them fact-side down on the table.

3 The Caller draws a Fact Triangle from the pile and extends the fact on the triangle by attaching one of the following:

- 1 zero to the dividend;
- 1 zero to the dividend and 1 zero to the divisor; or
- 2 zeros to the dividend.

4 The Caller reads the problem aloud and asks for the quotient.

5 The Calculator uses a calculator to solve the problem. The Brain solves it without a calculator. The Caller decides who got the answer first.

6 The Caller continues to draw one Fact Triangle at a time and ask for the quotient.

7 Players trade roles every 5 turns.

Example

The Caller draws the Fact Triangle with 3, 9, and 27. The Caller calls out an extended fact such as 270 / 9 or 270 / 30.

The Brain and Calculator each solve the problem.

The Caller decides who got the answer first.

Variations

- The Caller attaches 3 or more zeros to the dividend, or attaches 2 zeros to the dividend and 1 zero to the divisor.

- For a greater challenge, include Fact Triangles from *Math Masters*, page TA16.

Division Arrays

Materials	☐ number cards 6–18 (1 of each)
	☐ 1 six-sided die
	☐ 18 counters
Players	2 to 4
Skill	Modeling division with and without remainders
Object of the Game	To have the highest total score.

Directions

1 Shuffle the cards. Place the deck number-side down on the table.

2 Players take turns. When it is your turn, draw a card and take the number of counters shown on the card. You will use the counters to make an array.

- Roll the die. The number on the die is the number of equal rows you must have in your array.

- Make an array with the counters.

- Your score is the number of counters in one row. If there are no leftover counters, your score is double the number of counters in one row.

3 Players keep track of their scores. The player with the highest total score at the end of 5 rounds wins.

Example

Dave draws a 14 and takes 14 counters. He rolls a 3 and makes an array with 3 rows by putting 4 counters in each row. Two counters are left over.

Dave scores 4 for this round because there are 4 counters in each row.

Example

Marsha draws a 15 and takes 15 counters. She rolls a 3 and makes an array with 3 rows by putting 5 counters in each row.

Her score is $5 * 2 = 10$ for this round, because there are 5 counters in each row with none left over.

McGraw-Hill Education

Division Dash

Materials	☐ number cards 1–9 (4 of each)
	☐ 1 *Division Dash* Record Sheet for each player (*Math Masters*, p. G45)
Players	1 or 2
Skill	Dividing 2-digit numbers by 1-digit numbers
Object of the Game	To reach a score of 100.

Division Dash Record Sheet

	Division Problem	Quotient	Score
Sample	49 ÷ 4	12 R1	12
1			
2			
3			
4			
5			

Directions

1 Shuffle the cards and place the deck number-side down on the table.

2 Players take turns. When it is your turn:

- Turn over 2 cards and lay them on the table to make a 2-digit number. You may put the cards in any order. This is your *dividend*, the number you are dividing.

- Turn over another card. This is your *divisor*, the number you are dividing by.

- Divide the 2-digit number by the 1-digit number and record the result. This result is your *quotient*. Ignore any remainders. Calculate mentally or on paper.

- Add your quotient to your previous score and record your new score. If this is your first turn, your previous score was 0.

3 Players repeat Step 2 until one player's score is 100 or more. The first player to reach 100 or more wins. If there is only one player, the object of the game is to reach 100 in as few turns as possible.

Example

Turn 1: Curtis draws a 6, 4, and then 5. He divides 64 by 5.
He records the answer, 12 R4, on his record sheet.
The remainder is ignored. His score is $12 + 0 = 12$.

Turn 2: Curtis draws an 8, 2, and then 1. He divides 82 by 1.
He records the quotient, 82, on his record sheet. His score is $82 + 12 = 94$.

64 is the dividend.

5 is the divisor.

	Division Problem	Quotient	Score
Sample	49 ÷ 4	12 R1	12
1	64 ÷ 5	12 R4	12
2	82 ÷ 1	82	94

Factor Bingo

Materials
☐ number cards 2–10 (4 of each)

☐ 1 *Factor Bingo* Game Mat for each player (*Math Masters*, p. G19)

☐ 12 counters per player

Players 2 to 4

Skill Identifying factors of a number

Object of the Game To get 5 counters in a row, column, or diagonal; or to get 12 counters anywhere on the game mat.

Directions

1 Fill in your own game mat. Choose 25 different numbers from the numbers 2 through 90.

2 Write each number you choose in exactly 1 square on your game mat grid. The numbers should not all be in order, so be sure to mix them up as you write them on the grid. To help you keep track of the numbers you use, circle them in the list below the game mat.

3 Shuffle the cards and place them number-side down. A player turns over the top card. This card is the *factor*.

4 Players check their grids for a product that has the card number as a factor. Players who find a product cover it with a counter. A player may place only one counter on the grid for each card that is turned over.

5 Turn over the next card and repeat Step 4. The first player to get 5 counters in a row, column, or diagonal calls out "Bingo!" and wins the game. A player may also win by getting 12 counters anywhere on the game mat.

6 If all of the cards are used before someone wins, shuffle the cards again and continue playing.

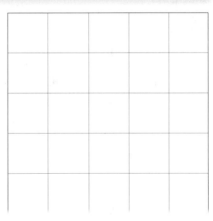

Factor Bingo Game Mat

```
    2   3   4   5   6   7   8   9  10
11  12  13  14  15  16  17  18  19  20
21  22  23  24  25  26  27  28  29  30
31  32  33  34  35  36  37  38  39  40
41  42  43  44  45  46  47  48  49  50
51  52  53  54  55  56  57  58  59  60
61  62  63  64  65  66  67  68  69  70
71  72  73  74  75  76  77  78  79  80
81  82  83  84  85  86  87  88  89  90
```

Numbers 2–90

Example

A 5 card is turned over. The number 5 is the *factor*. Any player may place a counter on one number that has 5 as a factor, such as 5, 10, 15, 20, or 25. A player may place only one counter on his or her game mat for each card that is turned over.

Factor Captor

Materials	☐ 1 *Factor Captor* Grid—either Grid 1 or Grid 2 (*Math Masters*, pp. G15–G16)
	☐ counters (48 for Grid 1; 70 for Grid 2)
	☐ 1 calculator for each player
Players	2
Skill	Finding factors of a number
Object of the Game	To have the higher total score.

Directions

1 To start the first round, Player 1 chooses a 2-digit number on the number grid, covers it with a counter, and records the number on scratch paper. This is Player 1's score for the round.

2 Player 2 covers all of the factors of Player 1's number. Player 2 finds the sum of the factors and records it on scratch paper. This is Player 2's score for the round.

A factor may only be covered once during a round.

3 If Player 2 missed any factors, Player 1 can cover them with counters and add them to his or her score.

4 In the next round, players switch roles. Player 2 chooses a number that is not covered by a counter. Player 1 covers all factors of that number.

5 Any number that is covered by a counter may not be used again.

6 The first player in a round may not cover a number that is less than 10, unless no other numbers are available.

7 Play continues with players trading roles after each round, until all numbers on the grid have been covered. Players then use their calculators to find their total scores. The player with the higher score wins the game.

1	2	2	2	2	2
2	3	3	3	3	3
3	4	4	4	4	5
5	5	5	6	6	7
7	8	8	9	9	10
10	11	12	13	14	15
16	18	20	21	22	24
25	26	27	28	30	32

Grid 1 (Beginning Level)

1	2	2	2	2	2	3
3	3	3	3	4	4	4
4	5	5	5	5	6	6
6	7	7	8	8	9	9
10	10	11	12	13	14	15
16	17	18	19	20	21	22
23	24	25	26	27	28	30
32	33	34	35	36	38	39
40	42	44	45	46	48	49
50	51	52	54	55	56	60

Grid 2 (Advanced Level)

Fishing for Digits

Materials	☐ 1 *Fishing for Digits* Record Sheet for each player (*Math Masters*, p. G7; optional)
	☐ 1 calculator for each player
Players	2
Skill	Understanding place value
Object of the Game	To have the larger number after 5 rounds.

Directions

1 Each player secretly enters a 6-digit number into his or her calculator. Zeros may not be used.

2 Player 1 goes "fishing" for a digit in Player 2's number by naming a digit.

3 If the digit named is one of the digits in Player 2's number:

- Player 2 reports the value of the digit. If the digit appears more than once in Player 2's number, Player 2 reports the largest value of that digit in the number. For example, if Player 1 names the digit 7, and Player 2's number is 987,675, then Player 2 would report the value 7,000, rather than the value 70.

- Player 1 *adds* the value of that digit to his or her number.

- Player 2 *subtracts* the value of that digit from his or her number.

4 If the digit named is not one of the digits in Player 2's number, Player 1 adds 0 and Player 2 subtracts 0 for that turn.

5 It is now Player 2's turn to "fish." Reverse the roles of Players 1 and 2 and repeat Steps 2, 3, and 4. When each player has "fished" once, the round is over.

6 The player whose calculator displays the larger number at the end of 5 rounds wins.

Example

Player 1's calculator shows 813296. Player 2's calculator shows 328479.

Player 1 asks: "Do you have the digit 4?"

Player 2 replies: "Yes. The value is 400."

Player 1 adds 400: 813296 ⊕ 400 (Enter) 813696.

Player 2 subtracts 400: 328479 ⊖ 400 (Enter) 328079.

Variation

Begin with a number having fewer than 6 digits.

Fishing for Fractions

Materials	☐ 1 set of Fraction Notation Cards (*Math Journal 1*, Activity Sheets 10–12)
	☐ 1 *Fishing for Fractions* Record Sheet for each player (*Math Masters*, p. G37)
Players	2 to 4
Skill	Adding fractions with like denominators
Object of the Game	To collect the most cards.

Directions

1 Shuffle the Fraction Notation Cards and deal 7 to each player. Place the remaining cards number-side down in the center of the table.

2 Players take turns. When it is your turn:

- Go "fishing" for a fraction in any other player's hand by asking for a card with a specific denominator (the same denominator as a card in your hand).

- If the other player has a card with that denominator, he or she will give it to you. Add the fractions on the two cards together. The other players check your answer. If you added correctly, write the equation on your record sheet and set the cards aside. If you added incorrectly, return the card to the other player.

- If the other player does not have a card with that denominator, go "fishing" by taking the top card from the deck. If you draw the denominator you asked for, add the cards, write the equation on your record sheet, and set the cards aside. If not, keep the new card in your hand. The next player starts a turn.

3 Play continues until someone runs out of cards or there are no cards left in the center. The winner is the person with the most cards set aside.

> **Example**
>
> Madison has a card with $\frac{4}{8}$ and asks if Jake has any eighths.
>
> Jake has $\frac{2}{8}$. He gives the card to Madison, who adds $\frac{4}{8} + \frac{2}{8} = \frac{6}{8}$. Madison records this equation on her record sheet and sets the cards aside.

Variations

Fishing for Fractions (Subtraction): The rules are the same as for *Fishing for Fractions*, except that players find the difference of the fractions instead of the sum. Remember to subtract the smaller fraction from the larger fraction.

Fishing for Fractions (Mixed-Number Addition): The rules are the same as for *Fishing for Fractions*, with two added steps:

- Shuffle a set of number cards (1–9 only) and place them in a pile separate from the Fraction Notation Cards.

- If you successfully fish for a fraction, either from another player or from the deck, draw two number cards. Use the Fraction Notation Cards and the number cards to make two mixed numbers on the *Fishing for Fractions* (Mixed-Number Addition) Gameboard (*Math Masters*, page G41). Add the mixed numbers.

Fishing for Fractions (Mixed-Number Subtraction): The rules are the same as for *Fishing for Fractions*, with two added steps:

- Shuffle a set of number cards (1–9 only) and place them in a pile separate from the Fraction Notation Cards.

- If you successfully fish for a fraction, either from another player or from the deck, draw two number cards. Use the Fraction Notation Cards and the number cards to make two mixed numbers on the *Fishing for Fractions* (Mixed-Number Subtraction) Gameboard (*Math Masters*, page G43), placing the larger mixed number on top. Subtract the mixed numbers.

Fraction/Decimal Concentration

Materials	☐ 1 set of *Fraction/Decimal Concentration* Cards (*Math Masters*, pp. G32–G33)
	☐ 1 calculator
Players	2 or 3
Skill	Recognizing fractions and decimals that are equivalent
Object of the Game	To collect the most cards by matching equivalent fraction and decimal cards.

Directions

Advance Preparation: Before beginning the game, cut out the cards and write the letter "F" on the back of each fraction card. Write the letter "D" on the back of each decimal card.

1 Create two separate sets of cards—a fraction set and a decimal set. Mix up the cards in each set and lay each set out separately, number-side down, so the "D" or "F" side is showing. Each set should be in 4 rows of 5 cards each.

2 Players take turns. On each turn, a player turns over both a fraction card and a decimal card. If the fraction and decimal are equivalent, the player keeps the cards. If the fraction and decimal are not equivalent, the player turns the cards back over so they are number-side down.

3 Players may use a calculator to check each other's matches.

4 The game ends when all of the cards have been taken. The player with the most cards wins.

Variation

Fraction Concentration: Create your own set of cards that are all fractions, so that each fraction card has exactly one equivalent fraction to match it. For example, $\frac{3}{5}$ and $\frac{6}{10}$ would be a match.

Fraction/Decimal Concentration Cards

$\frac{1}{10}$	$\frac{2}{10}$	$\frac{3}{10}$	$\frac{4}{10}$
$\frac{5}{10}$	$\frac{6}{10}$	$\frac{7}{10}$	$\frac{8}{10}$
$\frac{9}{10}$	$\frac{10}{10}$	$\frac{1}{100}$	$\frac{2}{100}$
$\frac{3}{100}$	$\frac{4}{100}$	$\frac{5}{100}$	$\frac{6}{100}$
$\frac{7}{100}$	$\frac{8}{100}$	$\frac{9}{100}$	$\frac{100}{100}$

0.1	0.2	0.3	0.4
0.5	0.6	0.7	0.8
0.9	1.0	0.01	0.02
0.03	0.04	0.05	0.06
0.07	0.08	0.09	1.00

Fraction Match

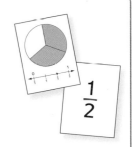

Materials	☐ 1 set of fraction cards
	☐ 1 set of WILD cards (*Math Journal 1*, Activity Sheet 13)
Players	2 to 4
Skill	Recognizing equivalent fractions
Object of the Game	To match all of your cards and have none left.

Directions

1 Shuffle the fraction cards and WILD cards together in one deck. Deal 7 cards to each player and place the remaining cards number-side down on the table. Turn over the top card and place it beside the deck. This is the *target card*. If a WILD card is drawn, return it to the bottom of the deck and continue drawing until the first target card is a fraction.

2 Players take turns trying to match the target card with a card from their hands in one of three possible ways:

- a card with an equivalent fraction,

- a card with a like denominator, or

- a WILD card: If a WILD card is played, the player names any fraction (with a denominator of 2, 3, 4, 5, 6, 8, 10, 12, or 100) that is equivalent to the target card.

Example

$\frac{1}{2}$ is the target card. It can be matched with:

- an equivalent fraction card, such as $\frac{2}{4}$, $\frac{3}{6}$, or $\frac{4}{8}$,

- a card with a like denominator, such as $\frac{0}{2}$, or

- a WILD card. The player can match $\frac{1}{2}$ by saying an equivalent fraction (but not $\frac{1}{2}$).

3 If a match is made, the player's matching card is placed on top of the pile and becomes the new target card. It is now the next player's turn. When a WILD card is played, the next player uses the fraction just stated for the new target card.

4 If no match is made, the player takes 1 card from the deck. If the card matches the target card, it may be played. If not, the player keeps the card and the turn ends.

5 The game is over when one of the players runs out of cards, when there are no cards left in the deck, or when time runs out. The player with the fewest cards wins.

Variation

Play with 1 set of Fraction Notation Cards (*Math Journal 1,* Activity Sheets 10–13).

Fraction Multiplication Top-It

Materials	☐	number cards 1–9 (4 of each)
	☐	1 set of Fraction Notation Cards (*Math Journal 1*, Activity Sheets 10–12)
	☐	1 *Top-It* Record Sheet for each player (*Math Masters*, p. G2)
	☐	fraction strips or fraction poster (optional)
Players	2	
Skill	Multiplying a whole number by a fraction	
Object of the Game	To collect more cards.	

Directions

1 Shuffle each deck of cards separately, and place them number-side down on the table.

2 Each player draws 1 number card and 1 Fraction Notation Card.

3 Each player multiplies his or her number card by his or her Fraction Notation Card.

4 Players compare products. The player with the larger product takes all the cards. Players may check their answers using any tool available. In the case of a tie, players repeat Steps 2 and 3.

5 Each player records both equations and the comparison on his or her *Top-It* Record Sheet.

6 The game ends when there are not enough cards left for each player to have another turn. The player with more cards wins.

Variations

• Use only number cards 1–5. Use fraction strips or another tool to multiply and compare numbers.

• Draw 2 number cards and 1 Fraction Notation Card to form an equation composed of a mixed number (1 number card and 1 Fraction Notation Card) multiplied by a whole number. For example, if a player draws the cards 3, 7, and $\frac{5}{8}$, he or she can write $3 * 7\frac{5}{8}$ or $7 * 3\frac{5}{8}$.

• Draw 2 number cards to form a 2-digit number to multiply by a fraction. For example, if a player draws the cards 3, 7, and $\frac{5}{8}$, he or she can write $37 * \frac{5}{8}$ or $73 * \frac{5}{8}$.

Fraction Top-It

Materials	☐ 1 set of fraction cards	
	☐ 1 *Top-It* Record Sheet for each player (*Math Masters*, p. G2)	
Players	2	
Skill	Comparing fractions	
Object of the Game	To collect more cards.	

$$\frac{5}{6} \qquad \frac{0}{4}$$

Directions

1 Deal 16 cards, number-side up, to each player.

2 Players spread their cards out, number-side up, so that all of the cards may be seen.

3 Players take turns, starting with the dealer. Each player plays one card. Place the cards number-side up in the center of the table.

4 Players compare cards. The player with the larger fraction wins the round and takes all of the cards. Players may check who has the larger fraction by turning over the cards and comparing the amounts shaded or the distance from 0 on the number line. Players should justify their comparisions. Each player records the comparison on his or her *Top-It* Record Sheet.

5 If the fractions are equivalent, each player plays another card. The player with the larger fraction takes all the cards from both plays.

6 The player who takes the cards starts the next round.

7 The game is over when all of the cards have been played. The player with more cards wins.

Variations

- Replace the fraction cards with one set of Fraction Notation Cards (*Math Journal 1*, Activity Sheets 10–12).

- To play with 3–4 players, deal 10 cards to each player for 3 players, or deal 8 cards to each player for 4 players. Each player plays one card. The player with the largest fraction takes all the cards. Players do not need to record their comparisons.

How Much More?

Materials	☐ 1 set of *How Much More?* Story Cards (*Math Masters*, p. G20)
	☐ 1 *How Much More?* Record Sheet for each player (*Math Masters*, p. G21)
	☐ 2 six-sided dice
Players	2
Skill	Solving comparison number stories
Object of the Game	To earn more points for creating valid comparisons.

Directions

1 Shuffle the *How Much More?* Story Cards and place the deck facedown.

2 Players take turns. When it is your turn:

- Draw a *How Much More?* Story Card, read the story aloud, and identify whether it is an additive comparison story or a multiplicative comparison story.

- Roll the dice. Use the sum you roll to fill in the first missing quantity in the story, and record this quantity in the second column on the record sheet.

- Roll the dice again and use the sum to determine the *number more than* or the *number of times more*. Record this number in the third column on the record sheet.

- Complete the problem on the card using the values rolled. Record your answer as the second quantity in the comparison. Read the comparison equation aloud.

- The other player checks your answer. If correct, you earn the *number more than* or *number of times more* value as points.

3 When all of the *How Much More?* Story Cards have been used, players add up their points. The player with more points wins.

Example

James drew the card shown at the right and read it aloud. James told the other player that it was a multiplicative comparison story. He first rolled 5 and 4, and found the sum of 9 as his first missing quantity. He then rolled 3 and 3, and found the sum of 6 as the *number of times more*. He completed the problem and found that Noah saved $54. He read the comparison equation, $9 * 6 = 54$, aloud and the other player confirmed that his work was correct. James earned 6 points for the round.

> Noah saved $ _____ last week. This week he saved _____ times as much money. How much money did Noah save this week?

Multiplication Wrestling

Materials	☐ 1 *Multiplication Wrestling* Record Sheet for each player (*Math Masters*, p. G35)
	☐ number cards 0–9 (4 of each) or 1 ten–sided die
Players	2
Skill	Multiplying 2-digit numbers using partial-products multiplication
Object of the Game	To get the larger product of two 2-digit numbers.

Directions

1 Shuffle the deck of cards and place it number-side down on the table.

2 Each player draws 4 cards and forms two 2-digit numbers. Players should form their numbers so their product is as large as possible.

3 Each player creates 2 "wrestling teams" by writing each of their numbers as a sum of 10s and 1s.

4 Each player's 2 teams wrestle. Each member of the first team (for example, 70 and 5) is multiplied by each member of the second team (for example, 80 and 4). Then the 4 products are added.

5 The player with the larger product wins the round and receives 1 point.

> ### Example
>
> | *Player 1:* | *Player 2:* |
> |---|---|
> | Draws 4, 5, 7, and 8. | Draws 1, 4, 9, and 6. |
> | Forms 75 and 84. | Forms 64 and 91. |
>
>
>
> | 75 * 84 | | 64 * 91 | |
> |---|---|---|---|
> | **Team 1** | **Team 2** | **Team 1** | **Team 2** |
> | (70 + 5) * (80 + 4) | | (60 + 4) * (90 + 1) | |
> | Products: | 70 * 80 = 5,600 | Products: | 60 * 90 = 5,400 |
> | | 70 * 4 = 280 | | 60 * 1 = 60 |
> | | 5 * 80 = 400 | | 4 * 90 = 360 |
> | | 5 * 4 = 20 | | 4 * 1 = 4 |
> | Total | 6,300 | Total | 5,824 |
> | 75 * 84 = 6,300 | | 64 * 91 = 5,824 | |
>
> 6,300 is greater than 5,824, so Player 1 gets 1 point.

6 To begin a new round, each player draws 4 new cards to form 2 new numbers. The player with more points at the end of 3 rounds is the winner.

Name That Number

Materials	☐ 1 set of number cards
Players	2 or 3
Skill	Finding equivalent names for numbers using operations
Object of the Game	To collect the most cards.

Directions

1 Shuffle the deck and deal 5 cards to each player. Place the remaining cards number-side down on the table between the players. Turn over the top card and place it beside the deck. This is the *target number* for the round.

2 Players try to match the target number by adding, subtracting, multiplying, or dividing the numbers on as many of their cards as possible. A card may only be used once.

3 Players write their solutions on a sheet of paper using grouping symbols as needed.

When players have written their best solutions:

- Each player sets aside the cards he or she used to match the target number.

- Each player replaces the cards he or she set aside by drawing new cards from the top of the deck.

- The old target number is placed on the bottom of the deck.

- A new target number is turned over, and another round is played.

4 Play continues until there are not enough cards left to replace all of the players' cards. The player who has set aside the most cards wins the game.

Example

Target number: 16

Player 1's cards:

Some possible solutions:

$10 + 8 - 2 = 16$ (3 cards used)

$10 + (7 * 2) - 8 = 16$ (4 cards used)

$10 / (5 * 2) + 8 + 7 = 16$ (all 5 cards used)

The player sets aside the cards used to make a solution and draws the same number of cards from the top of the deck.

Number Top-It

Materials	☐ number cards 0–9 (4 of each)
	☐ 1 *Top-It* Record Sheet for each player (*Math Masters*, p. G2)
	☐ 1 *Number Top-It* Mat (*Math Masters*, pp. G3–G4)
Players	2
Skill	Understanding place value for whole numbers
Object of the Game	To make the larger 6-digit number.

Directions

1 Shuffle the cards and place the deck number-side down on the table.

2 Each player uses one row of boxes on the *Number Top-It* Mat. In each round, players take turns turning over the top card from the deck and placing it number-side up on any one of their empty boxes. Each player takes a total of 6 turns, and places 6 cards on his or her row of the game mat.

3 At the end of each round, players read their numbers aloud and compare them. Each player records the comparison on his or her *Top-It* Record Sheet. The player with the larger number for the round scores 1 point. The other player scores 2 points.

4 Play 5 rounds for a game. Shuffle the deck between each round. The player with the smaller total number of points at the end of 5 rounds wins the game.

Example

Andy and Barb played *Number Top-It*. Here is the result of one complete round of play.

	Hundred-Thousands	Ten-Thousands	Thousands	Hundreds	Tens	Ones
Andy	6	4	5	2	0	1
Barb	9	7	3	5	2	4

Barb's number is larger, so Barb scores 1 point for this round, and Andy scores 2 points.

Variation

To play with 3–5 players, use 1 *Number Top-It* Mat (*Math Masters*, pages G3–G4) for every 2 players. Each player uses one row on a mat. Players take turns as above, then all players read and compare their numbers. The player with the largest number for the round scores 1 point, the player with the next-largest number scores 2 points, and so on.

Polygon Capture

Materials
☐ 1 set of *Polygon Capture* Pieces (*Math Masters*, p. G22)

☐ 1 set of *Polygon Capture* Property Cards (*Math Masters*, pp. G23–G24)

☐ 1 *Polygon Capture* Record Sheet for each player (*Math Masters*, p. G25)

Players 2, or 2 teams of two

Skill Identifying properties of polygons

Object of the Game To collect more polygons.

Directions

1 Spread the *Polygon Capture* Pieces out on the table. Shuffle the *Polygon Capture* Property Cards and sort them property-side down into "Angles" and "Sides" piles. (The cards are labeled on the back.)

There is only one right angle.	There are one or more right angles.	All angles are right angles.	There are no right angles.
There is at least one acute angle.	At least one angle is more than 90°.	All angles are right angles.	There are no right angles.
All opposite sides are parallel.	Only one pair of sides is parallel.	There are no parallel sides.	**Wild Card:** Pick your own side property.
At least two sides are perpendicular.	There are four perpendicular sides.	All the sides are the same length.	**Wild Card:** Pick your own side property.

Polygon Capture Property Cards
(property-side up)

2 Players take turns. When it is your turn:

- Draw the top card from each pile of property cards.

- Take all of the polygons that have **both** of the properties shown on the property cards you drew.

- If there are no polygons with both properties, draw one additional property card from either pile. Look for polygons that have this new property **and** one of the properties already drawn. Take these polygons.

- At the end of your turn, record the properties, the letters of the polygons captured, and the number of polygons captured on your record sheet.

- If you did not capture a polygon that you could have taken, the other player or team may name it, capture it, and add it to their score for the round.

3 When all the property cards in either pile have been drawn, shuffle the used property cards back into the deck. Sort the cards into "Angles" and "Sides" piles. Continue play.

4 The game ends after 5 rounds or when there are fewer than 3 polygons left.

5 Players add the number of polygons captured in each round to find their total. The winner is the player or team who has captured more polygons.

Example

Liz has these Property Cards: "All angles are right angles" and "All sides are the same length." She takes all the squares (Polygons A and H). Liz has "captured" these polygons.

Product Pile-Up

Materials   □   number cards 1–10 (8 of each)

Players   3 to 5

Skill   Practicing multiplication facts 1 to 10

Object of the Game   To play all of your cards and have none left.

Directions

1   Shuffle the cards and deal 12 cards to each player. Place the rest of the deck number-side down on the table.

2   The player to the left of the dealer begins. This player selects 2 of their cards, places them number-side up on the table, multiplies the numbers, and gives the product.

3   Play continues with each player selecting and playing 2 cards with a product that is *greater than* the product of the last 2 cards played.

> **Example**
>
> Joe plays 3 and 6 and says, "3 times 6 equals 18."
>
> The next player, Rachel, looks at her hand to find 2 cards with a product greater than 18. She plays 5 and 4 and says, "5 times 4 equals 20."

4   If a player is not able to play 2 cards with a greater product, the player must draw 2 cards from the deck. These 2 cards are added to the player's hand. If the player is now able to make a greater product, the 2 cards are played, and play continues.

5   If after drawing the 2 cards a player still cannot make a play, the player says "Pass." If all the other players say "Pass," the last player who was able to lay down 2 cards starts play again. That player may select any 2 cards to make *any* product and play continues.

6   If a player states an incorrect product, he or she must take back the 2 cards, draw 2 cards from the deck, and say "Pass." Play moves to the next person.

7   The winner is the first player to run out of cards, or the player with the fewest cards when there are no more cards to draw.

Rugs and Fences

Materials	☐ 1 set of *Rugs and Fences* Cards (*Math Masters*, p. G11)
	☐ 1 set of *Rugs and Fences* Rectangle Cards (*Math Masters*, pp. G12–G13)
	☐ 1 *Rugs and Fences* Record Sheet for each player (*Math Masters*, p. G14)
Players	2
Skill	Finding the area and perimeter of rectangles by applying formulas
Object of the Game	To score more points.

Directions

1 Shuffle the *Rugs and Fences* Rectangle Cards and place them picture-side down.

2 Shuffle the *Rugs and Fences* Cards and place them writing-side down next to the Rectangle Cards.

3 Players take turns. When it is your turn, draw one card from each deck and place the cards faceup on the table.

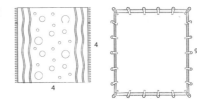

- If you draw an area (*A*) card, find the area.

- If you draw a perimeter (*P*) card, find the perimeter.

- If you draw a "Player's Choice" card, *you* may choose to find either the area or the perimeter using a formula.

- If a "Partner's Choice" card is drawn, your *partner* chooses whether you will find the area or the perimeter of the rectangle.

4 During your turn, record the rectangle's length and width and circle *A* (area) or *P* (perimeter) on your record sheet. Then write a number sentence to show how you used a formula to find the area or perimeter. The answer is your score for the round.

5 The player with the higher total score at the end of 6 rounds is the winner.

Example

Jonah draws the cards at the right. He may choose to calculate the area or the perimeter. Before he answers, Jonah figures out both the area and perimeter in his head.

Area = 3 * 4 = 12 square units

and

Perimeter = 14 units

Jonah records the length and width, and circles *P* on his record sheet. He writes the number model 2 * (3 + 4) = 14, and earns 14 points.

Spin-and-Round

Materials	☐ 1 *Spin-and-Round* Spinner (*Math Masters*, p. G6)
	☐ 1 *Spin-and-Round* Record Sheet (*Math Masters*, p. G5)
	☐ number cards 0–8 (4 of each)
	☐ 1 pencil and 1 large paper clip for the spinner
Players	2
Skill	Rounding 6-digit numbers to the nearest hundred through hundred-thousand
Object of the Game	To have the greater total score.

Directions

1 Shuffle the cards and place the deck number-side down.

2 Players take turns. When it is your turn:

- Draw 6 number cards to form a number. Place the first card faceup in the hundred-thousands place. Place the next card in the ten-thousands place, and so on, until you place the sixth card in the ones place. Record the starting number on the *Spin-and-Round* Record Sheet.

- Spin the spinner to determine how to round the number. On the record sheet, circle the digit to be rounded.

- Round the number. Record the rounded number on the record sheet.

- Your score for each turn is the place-value digit that was rounded up or that stayed the same. For example, if you round 482,657 to the nearest thousand, you get 483,000. Your score for this turn is 3 because the digit in the thousands place is 3. Record your score on the record sheet.

3 After 6 turns, each player finds the sum of their scores.

4 The player with the greater total score wins the game.

Variation

After spinning, players may rearrange their 6 cards before rounding. Keep in mind that the order of your numbers might help you score more points.

Subtraction Target Practice

Materials	☐ number cards 1–9 (4 of each)
	☐ 1 calculator for each player
Players	1 or more
Skill	Subtracting 2-digit numbers
Object of the Game	To get as close as possible to 0, without going below it.

Directions

1 Shuffle the cards and place the deck number-side down on the table. Each player starts at 250.

2 Players take turns. Each player has 5 turns in a game. When it is your turn, do the following:

- **Turn 1:** Turn over the top 2 cards and make a 2-digit number. (You may place the cards in either order.) Subtract this number from 250 on scratch paper. Check the answer on a calculator.

- **Turns 2–5:** Take 2 cards and make a 2-digit number. Subtract this number from the result obtained in your previous subtraction problem. Check the answer on a calculator.

3 The player whose final result is closest to 0, without going below 0, is the winner. If all players' final results are below 0, no one wins.

If there is only 1 player, the object of the game is to get as close to 0 as possible, without going below 0.

Example

Turn 1: Draw 4 and 5. Subtract 45 or 54. $250 - 45 = 205$

Turn 2: Draw 0 and 6. Subtract 6 or 60. $205 - 60 = 145$

Turn 3: Draw 4 and 1. Subtract 41 or 14. $145 - 41 = 104$

Turn 4: Draw 3 and 2. Subtract 32 or 23. $104 - 23 = 81$

Turn 5: Draw 6 and 8. Subtract 68 or 86. $81 - 68 = 13$

Each player has 5 turns in a game. This player's final result is 13.

Variation

Each player starts at 100 instead of 250.

Top-It Games

Materials	☐ number cards 1–9 (4 of each)
	☐ 1 *Top-It* Record Sheet for each player (*Math Masters*, p. G2)
Players	2
Skill	Adding, subtracting, multiplying, and dividing
Object of the Game	To collect more cards.

Addition Top-It (Advanced Version)

Directions

1 Shuffle the deck and place it number-side down on the table. Each player turns over 6 cards, forms two 3-digit numbers, and finds the sum of the numbers. Players should carefully consider how they form their numbers, since different arrangements have different sums. For example, 741 + 652 has a greater sum than 147 + 256.

2 Each player says his or her equation aloud. Players compare their sums and each player records both equations and the comparison on his or her record sheet.

3 The player with the larger sum takes all the cards. The game ends when there are not enough cards for each player to have another turn. The player with more cards wins.

Variation

• Each player turns over 8 cards, forms two 4-digit numbers, and finds the sum.

• To play with 3–4 players, all players compare their sums. The player with the largest sum takes all the cards. Players do not need to record their comparisons.

Subtraction Top-It (Advanced Version)

Directions

1 Shuffle the deck and place it number-side down. Each player turns over 6 cards, forms two 3-digit numbers, and finds the difference. Players should consider how they form their numbers. For example, 751 − 234 has a greater difference than 517 − 342.

2 Each player says his or her equation aloud. Players compare their differences and each player records both equations and the comparison on his or her record sheet.

3 The player with the larger difference takes all the cards. The game ends when there are not enough cards for each player to have another turn. The player with more cards wins.

Variation

To play with 3–4 players, the player with the largest difference takes all the cards.

Multiplication Top-It (Advanced Version)

Directions

1. Shuffle the deck and place it number-side down on the table.

2. Each player turns over 4 cards, uses them to form one 3-digit number and one 1-digit number, and then finds the product of the numbers.

3. Compare products. Each player records both players' equations and the comparison on his or her record sheet. The player with the larger product takes all the cards.

4. The game ends when there are not enough cards left for each player to have another turn. The player with more cards wins.

Variations

- Each player turns over 4 cards, forms two 2-digit numbers, and finds the product.

- Each player turns over 2 cards, attaches a zero to the first card drawn, and then multiplies by the number on the second card. For example, if 7 is the first card drawn and 5 is the second card drawn, find $70 * 5 = 350$.

- To play with 3–4 players, the player with the largest product takes all the cards. Players do not need to record their comparisons.

Division Top-It

Directions

1. Shuffle the deck and place it number-side down on the table.

2. Each player turns over 3 cards and uses them to generate a division problem as follows:
   - Choose 2 cards to form the dividend.
   - Use the remaining card as the divisor.
   - Divide and ignore any remainder.

3. Compare quotients. Each player records both players' equations and the comparison on his or her record sheet. The player with the larger quotient takes all the cards.

4. The game ends when there are not enough cards left for each player to have another turn. The player with more cards wins.

Variations

- Each player turns over 4 cards, chooses 3 of them to form a 3-digit number, then divides the 3-digit number by the remaining number. Players should carefully consider how they form their 3-digit numbers. For example, $462 / 5$ is greater than $256 / 4$.

- To play with 3–4 players, the player with the largest quotient takes all the cards.

Introduction

We use data as a part of everyday life. Scientists, historians, mathematicians, and others collect data. They collect data by counting, measuring, or observing and then organize their data to share with others. Analyzing data helps us answer questions about real-life problems or situations. This section provides you with a collection of interesting data from the world around you.

Some of the information you will see in this section could be easy to find on your own because it comes from a single source. For example, you can find the number of people who lived in different cities in the United States over time by looking at information from the U.S. Census, which is available online and in books. Some of the information comes from many different sources. In that case, it would take you a long time to collect the data on your own. For example, to find out how many animals live in zoos around the world, you would have to search the zoos' individual websites or find books about them. Instead, here the information has been collected and organized into tables and maps, so you can focus on using the data.

You might notice things about the data that make you curious. You can collect more information through measurement, observation, or research. Ask an adult to help you use tools such as a computer, the Internet, or reference books to do more research.

Pages 289–290 provide a list of the sources used to find the data in this section.

Lengths of Roller Coasters

Roller coasters can be very long in length or have very long drops. The following table gives information on some of the longest roller coasters around the world.

Roller Coasters			
Roller Coaster	Location	Length (in feet)	Longest Drop (in feet)
Kingda Ka	Six Flags Great Adventure, New Jersey, United States	3,118	418
The Ultimate	Lightwater Valley, North Yorkshire, England	7,450	Not available
Steel Dragon 2000	Nagashima Spa Land, Mie Prefecture, Japan	8,133	307
Goliath	Six Flags Magic Mountain, California, United States	4,480	255
Fujiyama	Fuji-Q Highland, Yamanashi Prefecture, Japan	6,709	230
Millennium Force	Cedar Point, Ohio, United States	6,595	300
Intimidator 305	Kings Dominion, Virginia, United States	5,100	300
The Desperado	Primm Valley Resorts, Nevada, United States	5,843	225

chaimdan/E+/Getty Images

Sizes of Indoor Water Parks

The following is a list of some indoor water parks around the world and their approximate areas.

Areas of Indoor Water Parks		
Indoor Water Park	Location	Area (in square feet)
Avalanche Bay	Boyne Falls, Michigan, United States	88,000
Beijing National Aquatics Center	Beijing, China	129,000
Great Wolf Lodge	Niagara Falls, Ontario, Canada	103,000
Kalahari Resorts	Sandusky, Ohio, United States	173,000
Kalahari Resorts	Wisconsin Dells, Wisconsin, United States	125,000
Sahara Sam's Oasis	West Berlin, New Jersey, United States	58,000
Tropical Islands Resort	Krausnick, Brandenburg, Germany	710,000
Wilderness at the Smokies	Sevierville, Tennessee, United States	75,000
World Waterpark	Edmonton, Alberta, Canada	225,000

Pavel Losevsky/iStock/Getty Images Plus/Getty Images

N

Sizes of Zoos around the World

There are many ways to describe the size of a zoo. How much land does it cover (what is its area)? How many animals can you see when you visit? How many different kinds (or species) of animals live at the zoo? The table below gives you this information for some of the largest zoos in the world.

Zoos around the World				
Name	Location	Approximate Area* (in square meters)	Number of Animals	Number of Species
Berlin Zoological Garden	Berlin, Germany	344,000	20,365	1,504
Bronx Zoo	Bronx, New York, United States	1,072,000	6,624	618
Columbus Zoo and Aquarium	Columbus, Ohio, United States	2,355,000	10,264	577
Denver Zoo	Denver, Colorado, United States	801,000	3,857	627
Omaha's Henry Doorly Zoo and Aquarium	Omaha, Nebraska, United States	526,000	17,000	962
London Zoo	London, England	150,000	2,264	736
Moscow Zoo	Moscow, Russia	214,000	7,755	1,127
National Zoological Park	Washington, D.C., United States	486,000	5,989	496
National Zoological Gardens of South Africa	Pretoria, Gauteng Province, South Africa	850,000	9,087	705
Toronto Zoo	Toronto, Ontario, Canada	2,873,000	6,086	486

*Areas have been rounded to the nearest 1,000 square meters.

Major U.S. City Populations in 2010

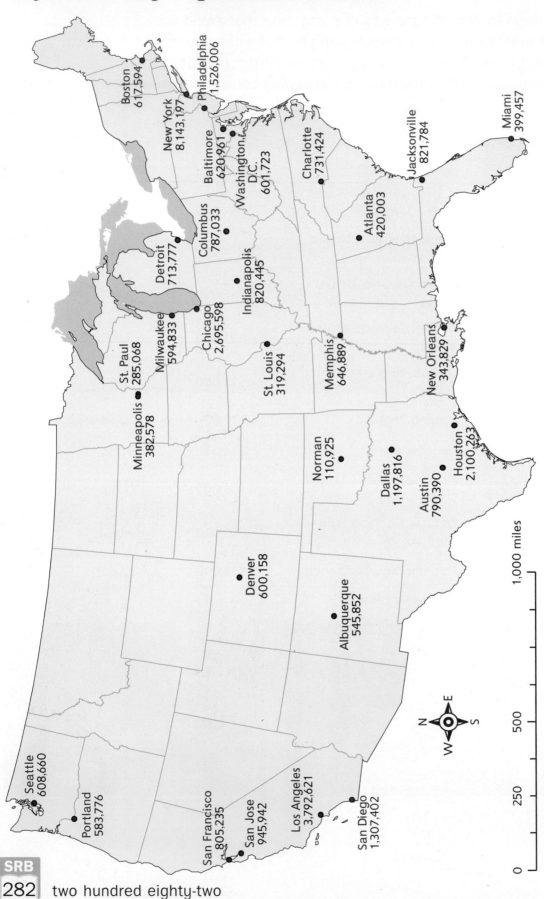

Boston
617,594

Philadelphia
1,526,006

New York
8,143,197

Baltimore
620,961

Washington,
D.C.
601,723

Charlotte
731,424

Jacksonville
821,784

Miami
399,457

Columbus
787,033

Atlanta
420,003

Detroit
713,777

Indianapolis
820,445

Milwaukee
594,833

Chicago
2,695,598

St. Louis
319,294

Memphis
646,889

New Orleans
343,829

St. Paul
285,068

Minneapolis
382,578

Norman
110,925

Dallas
1,197,816

Austin
790,390

Houston
2,100,263

Denver
600,158

Albuquerque
545,852

Seattle
608,660

Portland
583,776

San Francisco
805,235

San Jose
945,942

Los Angeles
3,792,621

San Diego
1,307,402

N
E
S
W

1,000 miles

500

250

0

Major U.S. City Populations in 1930

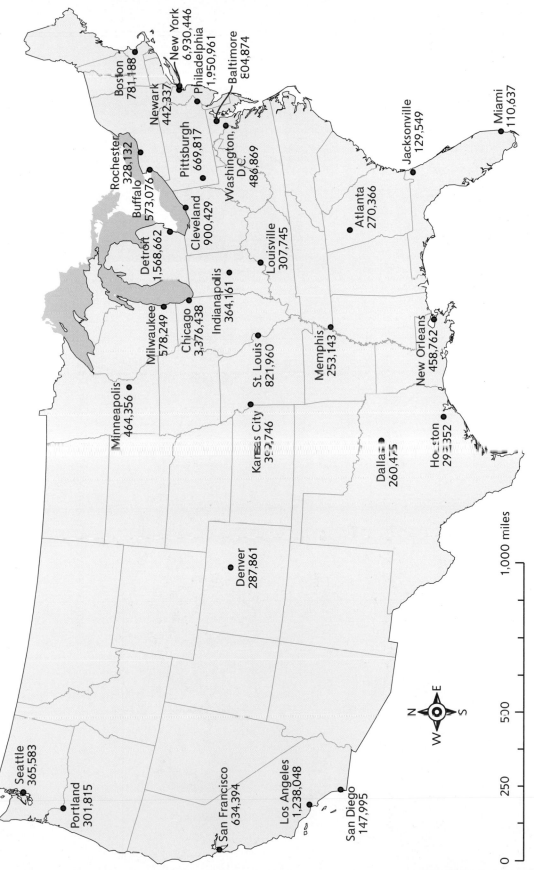

Boston 781,188
New York 6,930,446
Newark 442,337
Philadelphia 1,950,961
Baltimore 804,874
Rochester 328,132
Pittsburgh 669,817
Washington, D.C. 486,869
Jacksonville 129,549
Miami 110,637
Buffalo 573,076
Cleveland 900,429
Atlanta 270,366
Detroit 1,568,662
Louisville 307,745
Indianapolis 364,161
Milwaukee 578,249
Chicago 3,376,438
Memphis 253,143
New Orleans 458,762
Minneapolis 464,356
St. Louis 821,960
Kansas City 399,746
Dallas 260,475
Houston 292,352
Denver 287,861
Seattle 365,583
Portland 301,815
San Francisco 634,394
Los Angeles 1,238,048
San Diego 147,995

1,000 miles

N
E
S
W

0   250   500   1,000

Normal September Rainfall (in centimeters)

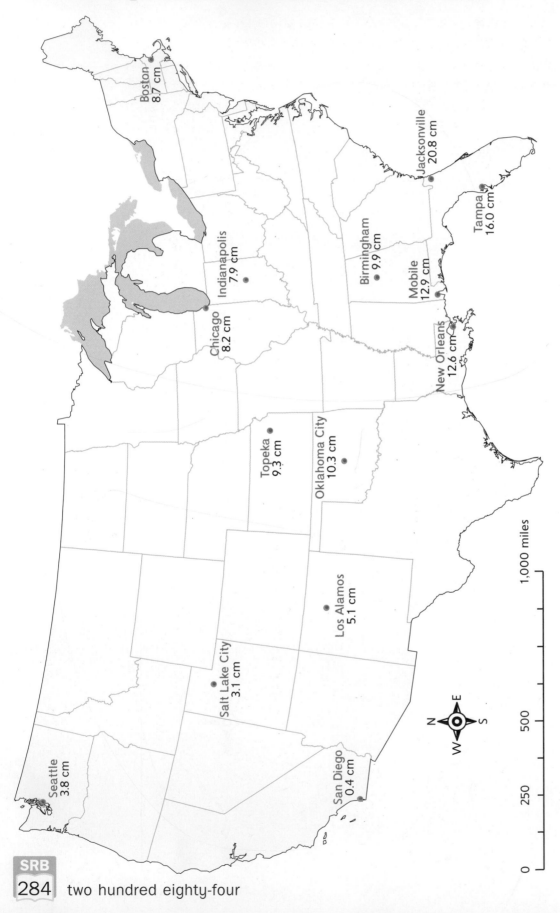

Boston
8.7 cm

Jacksonville
20.8 cm

Tampa
16.0 cm

Indianapolis
7.9 cm

Birmingham
9.9 cm

Mobile
12.9 cm

Chicago
8.2 cm

New Orleans
12.6 cm

Topeka
9.3 cm

Oklahoma City
10.3 cm

Los Alamos
5.1 cm

Salt Lake City
3.1 cm

Seattle
3.8 cm

San Diego
0.4 cm

Amounts given are averages for 30 years of data (1981–2010).

N
E
S
W

0   250   500   1,000 miles

Normal Monthly Precipitation (in centimeters)

Instead of only recording how much rain normally falls in a city in one month, sometimes meteorologists, or scientists who study the weather, record how much precipitation a city gets in a month. Precipitation is any kind of water that falls from the sky, including rain, snow, sleet, and hail. The following table shows the normal amount of precipitation for three cities in each month of the year. The amounts show how much precipitation normally falls in that month, when averaged over 30 years (1981–2010).

Normal Monthly Precipitation (in centimeters)			
Month	Miami, FL	Seattle, WA	Tucson, AZ
January	4.1	14.1	2.4
February	5.7	8.9	2.2
March	7.6	9.4	1.9
April	8.0	6.9	0.8
May	13.6	4.9	0.6
June	24.6	4.0	0.5
July	16.5	1.8	5.7
August	22.6	2.2	6.1
September	25.0	3.8	3.3
October	16.1	8.8	2.3
November	8.3	16.7	1.4
December	5.2	13.6	2.4

What Do Americans Eat?

The U.S. Department of Agriculture conducts surveys to find out how much food Americans eat on average. They ask a large number of people to keep lists of all the foods they eat over a period of several days. These lists are then used to estimate how much of each food was eaten during one year. Americans eat on average more than 2,000 pounds of food per year. This is about $5\frac{1}{2}$ pounds of food per day.

The most current results show that Americans eat or drink about the following average amounts of these foods in one year:

48	pounds of apples
25	pounds of bananas
5	pounds of broccoli
21	pounds of candy
12	pounds of carrots
33	pounds of cheese
249	eggs
16	pounds of fish
13	pounds of ice cream
362	cups of milk
23	cups of yogurt
46	slices of pizza
30	pounds of lettuce
68	quarts of popcorn

Food Supplies around the World

The table below shows the amounts of different kinds of food available to a typical person for a year in different countries around the world. Some foods are more plentiful in some countries than in others.

Food Supplies Around the World (in pounds per person)						
	Meat	Fish	Eggs	Milk	Vegetables	Fruits
Australia	267	56	16	507	211	207
Belgium	169	55	30	496	282	132
Brazil	205	23	19	332	119	307
Bolivia	148	4	12	100	71	153
Canada	203	49	26	481	251	285
Chad	27	11	1	48	17	20
China	127	72	41	69	732	179
Ecuador	120	17	17	438	56	378
Finland	164	78	20	871	194	208
Haiti	40	11	1	43	42	147
India	9	17	5	177	178	111
Japan	108	118	42	157	223	113
Mexico	134	24	39	245	114	222
Mongolia	162	1	4	299	106	64
Nicaragua	60	11	10	176	22	106
Nigeria	21	38	8	18	142	131
Philippines	76	72	9	32	141	255
Poland	167	26	21	436	285	120
Russia	147	49	34	383	242	151
Rwanda	14	4	1	41	120	375
South Africa	131	13	16	121	100	86
Spain	205	93	30	396	273	176
United States	259	48	31	566	249	214

Songbird Wing Lengths

The data set below shows the wing lengths of several wood thrushes, as measured by the scientists studying them. The scientists caught the birds in fine nets during spring migration. The measurements are shown to the nearest $\frac{1}{8}$ inch.

Wing Lengths of Wood Thrushes	
Bird	Wing Length (in inches)
1	$4\frac{1}{8}$
2	$4\frac{1}{4}$
3	4
4	4
5	4
6	$4\frac{3}{8}$
7	4
8	$4\frac{1}{4}$
9	$3\frac{3}{4}$
10	4
11	$4\frac{1}{4}$

BEZERGHEANU Mircea/iStock/360/Getty Images

Data Sources

The lists below give many of the sources for the data on the pages in this section. The information on some pages came from just one or two sources. On other pages the information came from many different sources. Often the information came from one or more sources and then was checked using different sources.

The data on these pages are not always what you might find if you looked them up yourself. Data may change as experts discover or collect new information. Or you may find the results of different sets of data from those shown here.

Lengths of Roller Coasters

Six Flags

Bennett, D. (1999). *Roller coaster: Wooden and steel coasters, twisters and corkscrews*. London: Aurum Press.

Sizes of Indoor Water Parks

Avalanche Bay Indoor Waterpark

Great Wolf Lodge

Kalahari Resorts and Conventions

Splash Lagoon

Tropical Islands Resort

Ultimate Waterpark

West Edmonton Mall

Wilderness at the Smokies

Sizes of Zoos around the World

colszoo.org

www.denverzoo.org

moscowzoo.ru

nzg.ac.za

www.tierpark-berlin.de

Toronto Zoo

www.waza.org

www.zsl.org

Major U.S. City Populations in 2010

census.gov

Major U.S. City Populations in 1930

census.gov

Normal September Rainfall (in centimeters)

Southeast Regional Climate Center (SERCC), University of North Carolina

Normal Monthly Precipitation (in centimeters)

ncdc.noaa.gov

What Do Americans Eat?

agday.org

agmrc.org

ers.usda.gov

nass.usda.gov

Food Supplies around the World

faostat3.fao.org

Songbird Wing Lengths

Julia Elizabeth Kelso, Utah State University

©Radius Images/Alamy

Place-Value Chart

billions	100 millions	10 millions	millions	100 thousands	10 thousands	thousands	hundreds	tens	ones	.	tenths	hundredths	thousandths
1,000 millions	100,000,000s	10,000,000s	1,000,000s	100,000s	10,000s	1,000s	100s	10s	1s	.	0.1s	0.01s	0.001s
10^9	10^8	10^7	10^6	10^5	10^4	10^3	10^2	10^1	10^0	.	10^{-1}	10^{-2}	10^{-3}

Rules for Order of Operations

1. Do operations inside parentheses or other grouping symbols first.
2. Calculate all expressions with exponents (e.g., $10^2 = 10 * 10 = 100$).
3. Multiply and divide in order, from left to right.
4. Add and subtract in order, from left to right.

Prefixes

uni- .... one	tera- .... trillion (10^{12})
bi- .... two	giga- .... billion (10^9)
tri- .... three	mega- .... million (10^6)
quad- .... four	kilo- .... thousand (10^3)
penta- .... five	hecto- .... hundred (10^2)
hexa- .... six	deca- .... ten (10^1)
hepta- .... seven	uni- .... one (10^0)
octa- .... eight	deci- .... tenth (10^{-1})
nona- .... nine	centi- .... hundredth (10^{-2})
deca- .... ten	milli- .... thousandth (10^{-3})
dodeca- .... twelve	micro- .... millionth (10^{-6})
iccsa- .... twenty	nano- .... billionth (10^{-9})

Multiplication and Division Table

*, /	1	2	3	4	5	6	7	8	9	10
1	1	2	3	4	5	6	7	8	9	10
2	2	4	6	8	10	12	14	16	18	20
3	3	6	9	12	15	18	21	24	27	30
4	4	8	12	16	20	24	28	32	36	40
5	5	10	15	20	25	30	35	40	45	50
6	6	12	18	24	30	36	42	48	54	60
7	7	14	21	28	35	42	49	56	63	70
8	8	16	24	32	40	48	56	64	72	80
9	9	18	27	36	45	54	63	72	81	90
10	10	20	30	40	50	60	70	80	90	100

The numbers on the diagonal are square numbers.

Metric System

Units of Length

1 kilometer (km) = 1,000 meters (m)
1 meter = 10 decimeters (dm)
= 100 centimeters (cm)
= 1,000 millimeters (mm)
1 decimeter = 10 centimeters
1 centimeter = 10 millimeters

Units of Capacity and Liquid Volume

1 kiloliter (kL) = 1,000 liters (L)
1 liter = 1,000 milliliters (mL)
1 cubic centimeter = 1 milliliter

Units of Mass and Weight

1 metric ton (t) = 1,000 kilograms (kg)
1 kilogram = 1,000 grams (g)
1 gram = 1,000 milligrams (mg)

U.S. Customary System

Units of Length

1 mile (mi) = 1,760 yards (yd)
= 5,280 feet (ft)
1 yard = 3 feet
= 36 inches (in.)
1 foot = 12 inches

Units of Capacity and Liquid Volume

1 gallon (gal) = 4 quarts (qt)
1 quart = 2 pints (pt)
1 pint = 2 cups (c)
1 cup = 8 fluid ounces (fl oz)
1 fluid ounce = 2 tablespoons (tbs)
1 tablespoon = 3 teaspoons (tsp)

Units of Mass and Weight

1 ton (T) = 2,000 pounds (lb)
1 pound = 16 ounces (oz)

Units of Time

1 century = 100 years
1 decade = 10 years
1 year (yr) = 12 months
= 52 weeks (plus one or two days)
= 365 days (366 days in a leap year)
1 month (mo) = 28, 29, 30, or 31 days
1 week (wk) = 7 days
1 day (d) = 24 hours
1 hour (hr) = 60 minutes
1 minute (min) = 60 seconds (sec)

Equivalent Fractions

Starting Fraction:	Multiply both the numerator and denominator by:				
	2	3	4	5	10
$\frac{1}{2}$	$\frac{2}{4}$	$\frac{3}{6}$	$\frac{4}{8}$	$\frac{5}{10}$	$\frac{10}{20}$
$\frac{3}{4}$	$\frac{6}{8}$	$\frac{9}{12}$	$\frac{12}{16}$	$\frac{15}{20}$	$\frac{30}{40}$
$\frac{5}{8}$	$\frac{10}{16}$	$\frac{15}{24}$	$\frac{20}{32}$	$\frac{25}{40}$	$\frac{50}{80}$

Formulas	Meaning of Variables
Rectangles	
• Perimeter: $p = 2l + 2w$, or $p = 2 * (l + w)$ • Area: $A = l * w$	p = perimeter; l = length; w = width; A = area
Squares	
• Perimeter: $p = 4 * s$ • Area: $A = s^2$	p = perimeter; s = length of one side; A = area
Regular Polygons	
• Perimeter: $p = n * s$	p = perimeter; n = number of sides; s = length of one side

About Calculators

You can use calculators for working with whole numbers, fractions, and decimals. As with any mathematical tool or strategy, you need to think about when and how to use a calculator. It can help you compute quickly and accurately when you have many problems to do in a short time. Calculators can help you solve problems with very large and very small numbers that may be hard to do mentally or with pencil and paper. Whenever you use a calculator, estimation should be part of your work. Always ask yourself if the answer in the display makes sense.

Calculator A

There are many different kinds of calculators. Four-function calculators do little more than add, subtract, multiply, and divide whole numbers and decimals. More advanced scientific calculators let you find powers and perform some operations with fractions. After elementary school, you may use graphic calculators that draw graphs, find data landmarks, and do even more complicated mathematics.

There are many calculators that work well with *Everyday Mathematics*. If the instructions in this book don't work for your calculator or the keys on your calculator are not explained here, you can refer to the directions that came with your calculator, look them up online, or ask your teacher for help.

The keystrokes in this section refer to the sample calculator shown here. The instructions in this section will refer to this calculator as Calculator A.

Basic Operations on a Calculator

Many handheld calculators use light cells for power. If you press the ON key and see nothing on the display, hold the front of the calculator toward a light or a sunny window for a moment and then press ON again.

Entering and Clearing

Pressing a key on a calculator is called *keying in,* or *entering.* In this book, calculator keys, except numbers and decimal points, are shown in rectangular boxes: ⊕, ⊜, and ⊗ and so on. A set of instructions for performing a calculation is called a *key sequence.*

The simplest key sequences turn the calculator on and enter or clear numbers or other characters. These keys are labeled on the calculator on the previous page and are summarized below.

Calculator A	
Key	Purpose
On/Off	Turn the display on.
Clear and On/Off at the same time	Clear the display and the short-term memory.
Clear	Clear only the display.
⇐	Clear the last digit.

Always clear both the display and the memory each time you turn on your calculator. It is also a good idea to clear both after you finish a problem and before you start another one.

Note It is important to take proper care of your calculator. Dropping it, leaving it in the sun, or other kinds of carelessness may break it, ruin it, or make it less reliable.

Fractions and Mixed Numbers on a Calculator

Some calculators let you enter, rewrite, and do operations with fractions. Once you know how to enter a fraction, you can add, subtract, multiply, or divide them just like whole numbers and decimals.

Entering Fractions and Mixed Numbers

Most calculators that let you enter fractions use similar key sequences. Always start by entering the **numerator.** Then press the key that tells the calculator to begin writing a fraction.

Example

Enter $\frac{5}{8}$ as a fraction in your calculator.

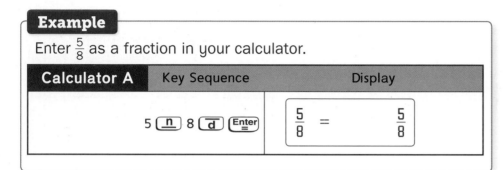

Note Pressing **d** after you enter the **denominator** is optional.

To enter a mixed number, enter the whole number and then press a key to tell the calculator this is the whole number part of a mixed number.

Example

Enter $73\frac{2}{5}$ as a fraction in your calculator.

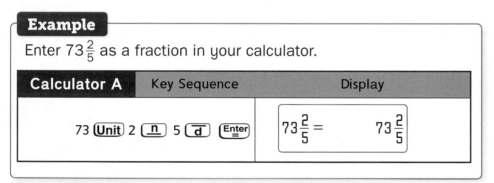

Try entering a mixed number in your calculator.

Skip Counting on a Calculator

To skip count on a calculator, you need to tell the calculator:

- what number to count by
- whether to count up or down
- what the starting number is
- when to count

Here's how to program Calculator A.

Op1 and Op2 allow you to program and repeat operations.

Example

Starting at 3, count by 7s.

Calculator A		
Purpose	Key Sequence	Display
Tell the calculator to count up by 7. **Op1** is programmed to do any operation with any number that you enter between presses of **Op1**.	**Op1** **+** 7 **Op1**	Op1 +7
Tell the calculator to start at 3 and do the first count.	3 **Op1**	Op1 3+7 1 10
Tell the calculator to count again.	**Op1**	↑ Op1 10+7 2 17
Keep counting by pressing **Op1**.	**Op1**	↑ Op1 17+7 3 24

To count back by 7, begin with **Op1** **−** 7 **Op1**.

Note You can use **Op2** to define a second counts-by operation. **Op2** works in exactly the same way as **Op1**.

Note The number in the lower left corner of the display shows how many counts you have made. When an arrow is in the display, you can scroll up or down to see the multiple steps in a problem.

Skip Counting by Fractions on a Calculator

You can also skip count by fractions on a calculator.

Example

Starting at $\frac{1}{2}$, count by $\frac{1}{4}$s.

Calculator A		
Purpose	Key Sequence	Display
Tell the calculator to count by $\frac{1}{4}$. **Op1** is programmed to do any operation with any number that you enter between presses of **Op1**.	**Op1** **+** 1 **n** 4 **d** **Op1**	Op1 $+\frac{1}{4}$
Tell the calculator to start at $\frac{1}{2}$ and do the first count.	1 **n** 2 **d** **Op1**	Op1 1 $\frac{3}{4}$
Tell the calculator to count again.	**Op1**	↑ Op1 2 1
Keep counting by pressing **Op1**.	**Op1**	↑ Op1 3 $1\frac{1}{4}$

To count back by $\frac{1}{4}$, begin with **Op1** **−** 1 **n** 4 **d** **Op1**.

Check Your Understanding

Program your calculator to do the following counts. Write five counts each.

1. Starting at 22, count up by 8s.

2. Starting at 146, count back by 16s.

3. Starting at $\frac{1}{3}$, count up by $\frac{2}{3}$.

Check your answers in the Answer Key.

Adding and Subtracting Fractions on a Calculator

You can compute with fractions and mixed numbers on a calculator.

Note See page 296 for instructions on entering fractions and mixed numbers on a calculator.

Example

Solve $3\frac{1}{8} + 1\frac{3}{8}$.

Calculator A	Key Sequence	Display
	3 (Unit) 1 (n) 8 (d) (+) 1 (Unit) 3 (n) 8 (d) (Enter)	$3\frac{1}{8} + 1\frac{3}{8} = \quad 4\frac{4}{8}$

You can multiply and divide fractions in similar ways by using the ⓧ and ÷ keys.

Example

Solve $\frac{9}{2} - \frac{7}{3}$.

Calculator A	Key Sequence	Display
	9 (n) 2 (d) (−) 7 (n) 3 (d) (Enter)	$\frac{9}{2} - \frac{7}{3} = \quad 2\frac{1}{6}$
	(U$\frac{n}{d}$⟷$\frac{n}{d}$)	$\frac{10}{6}$
	(U$\frac{n}{d}$⟷$\frac{n}{d}$)	$2\frac{1}{6}$

$\frac{9}{2} - \frac{7}{3} = 2\frac{1}{6} = \frac{13}{6}$

You can use the (U$\frac{n}{d}$⟷$\frac{n}{d}$) key to change the notation in the display.

Check Your Understanding

Use your calculator to solve the problems.

1. $2\frac{5}{6} - 1\frac{3}{6}$ **2.** $\frac{5}{6} + \frac{12}{8}$ **3.** $4\frac{1}{3} - \frac{5}{3}$ **4.** $\frac{6}{8} + 4\frac{3}{4}$

Check your answers in the Answer Key.

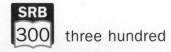

SRB
300 three hundred

0-9

2-dimensional (2-D) Having *area* but not volume. A 2-dimensional surface can be flat like a piece of paper or curved like a dome.

3-dimensional (3-D) Having length, width, and thickness. Solid objects take up *volume* and are 3-dimensional. A figure whose points are not all in a single plane is 3-dimensional.

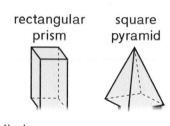

rectangular prism   square pyramid

cylinder   cone   sphere

A

accurate As correct as possible for the situation.

acute angle An angle with a measure less than 90°.

Acute angles

acute triangle A triangle with three acute angles.

Acute angles

addend Any one of a set of numbers that are added. For example, in 5 + 3 + 1 = 9, the addends are 5, 3, and 1.

additive comparison A situation involving two quantities and the difference between them.

adjacent angles *Angles* that are next to each other; adjacent angles have a common *vertex* and common *side,* but no other overlap. In the diagram, angles 1 and 2 are adjacent angles; so are angles 2 and 3, angles 3 and 4, and angles 4 and 1.

 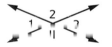

adjacent sides Two sides of a *polygon* with a common *vertex.*

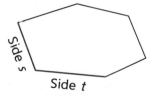

Sides *s* and *t* are adjacent sides of the hexagon.

algebra The branch of mathematics that uses letters and symbols to stand for *unknowns* and numbers that vary. Algebra is used to *model patterns,* numerical relationships, and real-world situations.

algorithm A set of step-by-step instructions for doing something, such as carrying out a computation or solving a problem.

analog clock A clock that shows the time by the positions of the hour and minute hands.

Analog clock

angle A figure that is formed by two rays or line segments with a common endpoint. The rays or segments are called the sides of the angle. The common endpoint is called the *vertex* of the angle. Angles are measured in *degrees* (°).

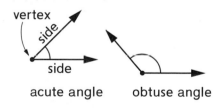

acute angle   obtuse angle

straight angle

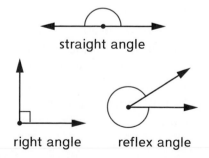

right angle   reflex angle

McGraw-Hill Education

approximate Close to exact. It is sometimes not possible to get an exact answer, but it is important to be close to the exact answer.

arc of a circle A part of a circle, for example, a semicircle is an arc.

Arcs

area The amount of *surface* inside a *2-dimesional* shape. The measure of the area is how many units, such as square inches or square centimeters, cover the surface.

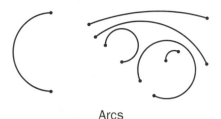

40 square units   about 21 square units

1 square centimeter   1 square inch

area model A *model* for multiplication problems in which the length and width of a *rectangle* represent the *factors*, and the *area* of the rectangle represents the *product*.

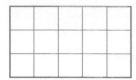

Area model for 3 * 5 = 15

array An arrangement of objects in a regular *pattern*, usually in *rows* and *columns*. In *Everyday Mathematics*, an array is a rectangular array unless specified otherwise.

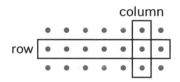

A rectangular array

Associative Property of Addition A *property* of addition (but not subtraction) that says that when you add three numbers, you can change the grouping without changing the *sum*. For example:
(4 + 3) + 7 = 4 + (3 + 7)

Associative Property of Multiplication A *property* of multiplication (but not division) that says that when you multiply three numbers, you can change the grouping without changing the *product*. For example:
(5 * 8) * 9 = 5 * (8 * 9)

ballpark estimate A rough *estimate* to help you solve a problem, check an answer, or when an exact answer cannot be found.

bar graph A graph that uses horizontal or vertical bars to represent *data*.

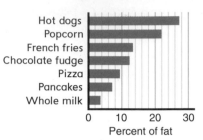

Fat Content of Foods

base ten Our system for writing numbers that uses 10 symbols called *digits*. The digits are 0, 1, 2, 3, 4, 5, 6, 7, 8, and 9. You can write any number using only these 10 digits. Each digit has a value that depends on its place in the number. In this system, moving a digit one place to the left makes that digit worth 10 times as much. And moving a digit one place to the right makes that digit worth one-tenth as much. See *place value*.

basic facts The addition facts (whole-number addends of 10 or less) and their related subtraction facts, and the multiplication facts (whole number factors of 10 or less) and their related division facts. Facts are organized into *fact families*.

benchmark A well-known number or measure that can be used to check whether other numbers, measures, or estimates make sense. For example, a benchmark for length is that the width of a man's thumb is about one inch. The numbers 0, $\frac{1}{2}$, 1, $1\frac{1}{2}$ may be useful benchmarks for fractions.

break apart A multiplication fact strategy where you decompose a factor into smaller numbers. For example, 7 × 6 can be broken apart into 2 × 6 and 5 × 6 to make solving easier.

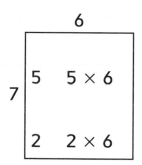

capacity (1) The amount a container can hold. Capacity is usually measured in units such as cups, fluid ounces, and liters. (2) The amount something can hold. For example, a computer hard drive may have a capacity of 64TB, or a scale may have a capacity of 400 lbs.

circle A 2-dimensional, closed, curved path whose points are all the same distance from a center point.

Circle

clockwise rotation A turning in the same direction as that of the hands of a clock.

close-but-easier numbers Numbers that are close to the original numbers in the problem, but easier for solving problems. For example, to estimate 494 + 78, you might use the close-but-easier numbers 480 and 80.

close-to estimation Estimation that uses close-but-easier numbers.

column A vertical ("up and down") arrangement of object in an *array*.

column

column addition A method for adding numbers in which the *addends'* digits are first added in each *place-value* column separately, and then 10-for-1 trades are made until each column has only one digit. Lines are drawn to separate the place-value columns.

100s	10s	1s
2	4	8
+ 1	8	7
3	12	15
3	13	5
4	3	5

248 + 187 = 435

common denominator For two or more fractions, a number that is a multiple of both or all *denominators*. For example, the fractions $\frac{1}{2}$ and $\frac{2}{3}$ have the common denominators 6, 12, 18, and so on. If the fractions have the same denominator, that denominator is called a common denominator.

common multiple A number that is a multiple of two or more given numbers. For example, common multiples of 6 and 8 include 24, 48, and 72.

common numerator Same as *like numerator*.

Commutative Property of Addition A *property* of addition (but not of subtraction) that says that changing the order of the numbers being added does not change the *sum*. This property is often called the *turn-around rule* in *Everyday Mathematics.* For example: $5 + 10 = 10 + 5$

Commutative Property of Multiplication A *property* of multiplication (but not of division) that says that changing the order of the numbers being multiplied does not change the *product*. This property is often called the *turn-around rule* in *Everyday Mathematics.* For example: $3 * 8 = 8 * 3$

complementary angles Two angles whose measures total 90°.

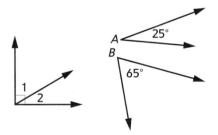

∠1 and ∠2; ∠A and ∠B are pairs of complimentary angles.

compose To make a number or shape by putting together smaller numbers or shapes. For example, you can compose a 10 by putting together ten 1s: $1 + 1 + 1 + 1 + 1 + 1 + 1 + 1 + 1 + 1 = 10.$ You can compose a pentagon by putting together an equilateral triangle and a square.

A composed pentagon

composite number A *counting number* that has more than two different *factors*. For example, 4 is a composite number because it has three factors: 1, 2, and 4.

composite unit A *unit* of measure made up of smaller units. For example, a foot is a composite unit of 12 inches, and a row of unit squares can be used to measure area.

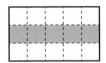

3 rows of 5 square units each have a total area of 15 square units.

conjecture A statement that is thought to be true based on information or mathematical thinking.

counterclockwise rotation A turning in the opposite direction as that of the hands of a clock.

counting numbers The numbers used to count things. The set of counting numbers is {1, 2, 3, 4, . . .}. Compare *whole numbers*.

counting-up subtraction A subtraction strategy in which you count up from the smaller to the larger number to find the *difference*. For example, to solve $16 - 9$, count up from 9 to 16.

D

data Information that is gathered by counting, measuring, questioning, or observing.

decimal A number written in standard, *base-10* notation that contains a *decimal point*, such as 2.54. A whole number is a decimal, but it is usually written without a decimal point.

decimal point A dot used to separate the ones and tenths places in *decimals*.

decompose To separate a number or shape into smaller numbers or shapes. For example, you can decompose 14 into 1 ten and 4 ones. You can decompose a square into two isosceles right triangles.

degree (°) (1) A unit of measure for *angles* based on dividing a *circle* into 360 equal parts. Latitude and longitude are measured in degrees, and these degrees are based on angle measures. (2) A unit of measure for *temperature*. A small raised circle (°) can be used to show degrees, as in a 70° angle or 70°F for room temperature.

denominator The number below the line in a *fraction*. A fraction may be used to name part of a *whole*. If the whole is divided into equal parts, the denominator represents the number of equal parts into which the whole is divided. The denominator determines the size of each part. For example, in $\frac{3}{4}$, 4 is the denominator.

difference The result of subtracting one number from another.

digit One of the number symbols 0, 1, 2, 3, 4, 5, 6, 7, 8, and 9 in the standard, *base-ten* system.

digital clock A clock that shows the time with numbers of hours and minutes, usually separated by a colon. For example, this digital clock shows three o'clock.

A digital clock

Distributive Property of Multiplication over Addition and Subtraction A *property* that relates multiplication and addition or subtraction. This property gets its name because it "distributes" a *factor* over terms inside parentheses. For example:

$$2 * (5 + 3) = (2 * 5) + (2 * 3)$$
$$= 10 + 6 = 16$$

and

$$2 * (5 - 3) = (2 * 5) - (2 * 3)$$
$$= 10 - 6 = 4$$

dividend The number in division that is being divided. For example, in $35 \div 5 = 7$, the dividend is 35.

divisible by If one *counting number* can be divided by a second counting number with a *remainder* of 0, then the first number is divisible by the second number. For example, 28 is divisible by 7 because 28 divided by 7 is 4, with a remainder of 0.

divisor In division, the number that divides another number. For example, in $35 \div 5 = 7$, the divisor is 5.

edge Any *side* of a *polyhedron's faces*.

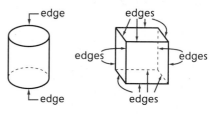

efficient strategy A method that can be applied easily and quickly.

elapsed time An amount of time that has passed. For example, between 12:45 P.M. and 1:30 P.M., 45 minutes have elapsed.

endpoint A point at the end of a line segment, ray, or curve. A line segment is named using the letter labels of its endpoints. A ray is named using the letter labels of its endpoint and another point on the ray.

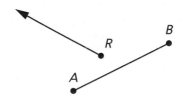

R, A, and B are endpoints.

equal See *equivalent*.

equal parts *Equivalent* parts of a *whole*. For example, dividing a pizza into 4 equal parts means each part is $\frac{1}{4}$ of the pizza and is equal in size to each of the other 3 parts.

equation A *number sentence* that contains an equal sign. For example, $15 = 10 + 5$ is an equation.

equilateral triangle A *triangle* with all three sides equal in length. In an equilateral triangle, all three angles have the same measure.

An equilateral triangle

equivalent *Equal* in value but possibly in a different form. For example, $\frac{1}{2}$, 0.5, and 50% are all equivalent.

equivalent fractions *Fractions* that name the same number. For example, $\frac{1}{2}$ and $\frac{4}{8}$ are equivalent fractions.

equivalent fractions rule A rule stating that if the *numerator* and *denominator* of a *fraction* are each multiplied or divided by the same nonzero number, the result is a fraction *equivalent* to the original fraction.

estimate An answer close to an exact answer. To estimate means to give an answer that should be close to an exact answer.

evaluate a numerical expression To carry out the *operations* in a numerical *expression* to find a single value for the expression.

even number A *counting number* that can be divided by 2 with no remainder. The even numbers are 2, 4, 6, 8, and so on.

expanded form A way of writing a number as the *sum* of the values of each *digit*. For example, in expanded form, 356 is written $300 + 50 + 6$. Compare *standard form*.

expression A group of mathematical symbols that represents a number—or can represent a number if values are assigned to any *variables* in the expression. An expression may include numbers, variables, *operation symbols*, and *grouping symbols*—but *not relation symbols* $(=, >, <,$ and so on). For example: 5, $6 + 3$, $(16 \div 2) - 5$, and $3 * m + 1.5$ are expressions.

extended facts Variations of basic facts involving multiples of 10, 100, and so on. For example, $30 + 70 = 100$, $40 * 5 = 200$, and $560/7 = 80$ are extended facts.

face A flat *surface* on the outside of a solid.

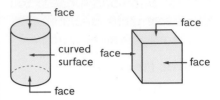

fact family A set of related addition and subtraction facts, or related multiplication and division facts. For example, $5 + 6 = 11$, $6 + 5 = 11$, $11 - 5 = 6$, and $11 - 6 = 5$ are a fact family. $5 * 7 = 35$, $7 * 5 = 35$, $35 \div 5 = 7$, and $35 \div 7 = 5$ are another fact family.

factor Whenever two or more numbers are multiplied to give a *product*, each of the numbers that is multiplied is called a factor. For example, in $4 * 1.5 = 6$, 6 is the product and 4 and 1.5 are called factors. Compare *factor of a counting number* n.

factor of a counting number *n* A *counting number* whose product with another counting number equals *n*. For example, 2 and 3 are *factors* of 6 because $2 * 3 = 6$. But 4 is not a factor of 6 because $4 * 1.5 = 6$ and 1.5 is not a counting number.

factor pair Two *factors* of a *counting number* whose *product* is the number. A number may have more than one factor pair. For example, the factor pairs for 18 are 1 and 18, 2 and 9, and 3 and 6.

false number sentence A *number sentence* that is not true. For example, 8 = 5 + 5 is a false number sentence.

fluid ounce (fl oz) A U.S. customary unit of liquid *volume* equal to $\frac{1}{16}$ of a pint, or about 30 milliliters.

formula A general rule for finding the value of something. A formula is often written using letters, called *variables*, which stand for the quantities involved. For example, the formula for the area of a rectangle may be written as $A = l * w$, where *A* represents the area of the rectangle, *l* represents its length, and *w* represents its width.

fraction A number in the form $\frac{a}{b}$ or *a/b*. Fractions can be used to name part of a whole or part of a collection. A fraction may also be used to represent division. For example, $\frac{2}{3}$ can be thought of as 2 divided by 3. See *numerator* and *denominator*.

Fraction Circle Pieces A set of colored circles each divided into equal-size slices, used to represent *fractions*.

Frames and Arrows A diagram used in *Everyday Mathematics* to show a number pattern or sequence.

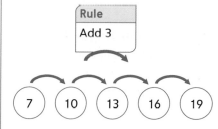

friendly numbers A *strategy* that uses numbers that are easy to work from, usually a multiple of 10. For example, to solve 16 − 9, you might recognize that the friendly number 10 is 6 less than 16, then count down 1 more to 9 to find that the difference is 7.

front-end estimation An estimation method that keeps only the left-most digit in the numbers and puts 0s in for all others. For example, the front-end *estimate* for 45,600 + 53,450 is 40,000 + 50,000 = 90,000.

function machine An imaginary machine that uses a rule to pair input numbers put in (inputs) with numbers put out (outputs). Each input is paired with exactly one output. Function machines are used in "What's My Rule?" problems.

G

geometric solid A *3-dimensional* shape, such as a *prism*, *pyramid*, cylinder, cone, or sphere. Despite its name, a geometric solid is hollow; it does not contain the points in its interior.

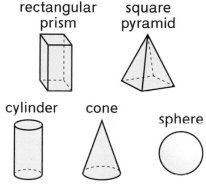

Geometry Template An *Everyday Mathematics* tool that includes a millimeter ruler, a ruler with sixteenth-inch intervals, half-circle and full-circle *protractors*, a percent circle, pattern-block shapes, and other geometric figures. The template can also be used as a compass.

grouping symbols Symbols such as parentheses ( ), brackets [ ], and braces { } that tell the order in which operations in an *expression* are to be done. For example, in the expression (3 + 4) * 5, the operation in the parentheses should be done first. The expression then becomes 7 * 5 = 35.

H

height (1) The length of the shortest line segment joining a corner of a shape to the line containing the base opposite it. (2) The line segment itself.

Heights/altitudes of 2-D figures are shown in blue.

Heights/altitudes of 2-D figures are shown in blue.

I

inequality A *number sentence* with >, <, ≥, ≤, or ≠. For example, the sentence 8 < 15 is an inequality.

intersect To meet or cross.

Intersecting segments

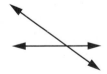

Intersecting lines

interval The set of all numbers between two numbers, *a* and *b*, which may include *a* or *b* or both.

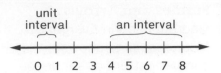

isosceles triangle A *triangle* with at least two *sides* equal in length. In an isosceles triangle, at least two *angles* have the same measure. A triangle with all three sides the same length is an isosceles triangle, but is usually called an equilateral triangle.

Isosceles triangles

iterate units To repeat a *unit* without gaps or overlaps in order to measure. For example, you can cover a surface by iterating or tiling a unit to measure *area*.

K

kilogram A metric unit of *mass* equal to 1,000 grams. A bottle of water is usually 1 kilogram.

kite A *quadrilateral* with two pairs of adjacent, equal-length *sides*. The four sides can all have the same length, so a rhombus is a kite.

Kites

L

lattice multiplication An old way to multiply multidigit numbers in a diagram that looks like a lattice.

length The distance between two points along a path.

like Equal or the same.

like denominator Same as *common denominator*.

like numerator A number that is the numerator of two or more fractions. For example, the fractions $\frac{3}{11}$ and $\frac{3}{7}$ have *common numerator* of 3.

line A straight path that goes on forever in both directions.

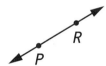

Line *RP* or *PR*

line of symmetry A line drawn through a figure so that it is divided into two parts that are mirror images of each other. The two parts look alike but face in opposite directions. See *line symmetry*.

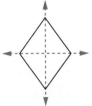 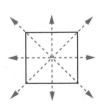

Lines of symmetry are shown in blue.

line plot A sketch of *data* in which check marks, Xs, or other marks above a labeled line show how many times each value appears in the set of data.

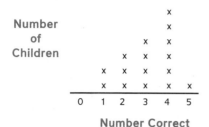

Test Scores

Number of Children

0 1 2 3 4 5

Number Correct

line segment A straight path joining two points. The two points are called *endpoints* of the segment.

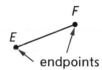

endpoints

Line segment EF or FE

line symmetry A figure has line symmetry if a line can be drawn through it so that it is divided into two parts that are mirror images of each other. The two parts look alike but face in opposite directions. See *line of symmetry*.

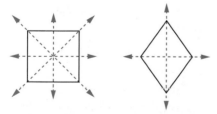

Lines of symmetry are shown in blue.

liquid volume An amount of liquid measured in units

such as liters and gallons. Units of liquid *volume* are frequently used to measure *capacity*.

liter (L) A metric unit of *volume* or *capacity* equal to the volume of a cube with 10 cm-long edges. A liter is a little larger than a quart.

M

mass A measure of how much matter is in object. Mass is often measured in grams or kilograms.

mathematical argument An explanation of why a claim is true or false using words, pictures, symbols, or other representations. For example, if you claim that $\frac{2}{5} > \frac{4}{7}$ is not true, you say that $\frac{2}{5}$ is less than $\frac{1}{2}$ and $\frac{4}{7}$ is more than $\frac{1}{2}$, so $\frac{2}{5}$ cannot be more than $\frac{4}{7}$.

mathematical practices Ways of working with mathematics. Mathematical practices are habits or actions that help people use mathematics to solve problems.

mathematical structure A relationship among mathematical objects, operations, or relations; a mathematical *pattern*, category, or *property*. For example, the Distributive Property of Multiplication over Addition is a structure of

arithmetic. The number grid illustrates some patterns and structures that exist in our number system.

measurement scale The spacing of the marks on a measuring device. The scales on this ruler are 1 millimeter on the left side and $\frac{1}{16}$ inch on the right side.

Scale of a number line

meter (m) The basic metric unit of length. A meter is about 39 inches, a little longer than a yard.

metric system A measurement system based on decimals and multiples of 10. The metric system is used by scientists and people in most countries around the world except the United States.

mixed number A number that is written using both a *whole number* and a *fraction*. For example, $2\frac{1}{4}$ is a mixed number equal to $2 + \frac{1}{4}$.

model A representation of a real-world object or situation. Number sentences, diagrams, and pictures can be models.

multiple of a number *n* A *product* of *n* and a *counting number*. For example, the multiples of 7 are 7, 14, 21, 28, and so on.

multiplicative comparison Statements about quantities that use multiplication as the comparison. For example, "Bev has three times as much money as Biff" is a multiplicative comparison.

multiplicative identity The number 1. The multiplicative identity is the number that when multiplied by any other number is that other number.

 N

name-collection box In *Everyday Mathematics*, a place to write *equivalent names* for a number.

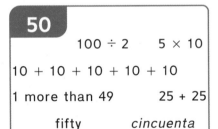

50
100 ÷ 2   5 × 10
10 + 10 + 10 + 10 + 10
1 more than 49   25 + 25
fifty   *cincuenta*

negative numbers A number that is less than zero; a number to the left of zero on a horizontal *number line* or below zero on a vertical number line. The symbol − may be used to write a negative number. For example, "negative 5" is usually written as −5.

number line A *line* with numbers marked in order on it.

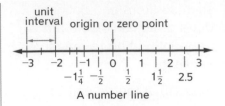

A number line

number model A group of numbers and symbols that models or fits a *number story* or situation. For example, the story *Sally had $5, and then she earned $8* can be modeled as 5 + 8 = 13 (a *number sentence*), or as 5 + 8 (part of a number sentence).

number sentence Two groups of mathematical symbols connected by a *relation symbol* (=, >, <, ≠). Mathematical symbols on each side of the number sentence can include numbers, letters for unknown values and/or operation symbols (+, −, *, or ÷). Number sentences often contain grouping symbols like parentheses. For example, 2 * (3 + 4) − 6 < 30 ÷ 3 and 4 * m = 100 are number sentences.

number story A story with a problem that can be solved using arithmetic.

numerator The number above the line in a *fraction*. A fraction may be used to name part of a *whole*. If the whole is divided into equal parts, the numerator represents the number of equal parts being considered. For example, in $\frac{3}{4}$, 3 is the numerator.

 O

obtuse angle An angle with measure greater than 90° and less than 180°.

Obtuse angles

obtuse triangle A triangle with an obtuse angle.

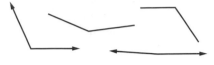

An obtuse triangle

open sentence A *number sentence* which has variables or missing numbers and is neither true nor false. For example, 5 + x = 13 is an open sentence. The sentence is true if 8 is substituted for x. The sentence is false if 4 is substituted for x.

operation An action performed on numbers or *expressions* to produce other numbers or expressions. Addition, subtraction, multiplication, and division are the four basic arithmetic operations.

operation symbol A symbol used to stand for a mathematical operation. Common operation symbols are +, − , *, ÷, and /.

order of operations Rules that tell in what order to perform operations in arithmetic.

ounce (oz) A U.S. customary unit equal to $\frac{1}{16}$ of a pound.

pan balance A tool used to weigh objects or compare *weights*.

parallel Always the same distance apart, and never meeting or crossing each other, no matter how far extended. Line segments are parallel if they are parts of lines that are parallel. The bases of a prism are parallel.

parallel bases

parallelogram A 4-sided polygon whose opposite sides are *parallel*. The opposite sides of a parallelogram are also the same length. And the opposite angles in a parallelogram have the same measure. All parallelograms are *trapezoids*.

Parallelogram

parentheses *Grouping symbols*, ( ), used to tell which parts of an expression should be calculated first.

partial-products multiplication (1) A way to multiply in which the value of each digit in one factor is multiplied by the value of each digit in the other factor. The final *product* is the sum of these partial products. (2) A similar method for multiplying mixed numbers.

partial-quotients division A way to divide in which the dividend is divided in a series of steps. The *quotients* for each step (called partial quotients) are added to give the final answer.

partial-sums addition A way to add in which sums are computed for each place (ones, tens, hundreds, and so on) separately. The partial-sums are then added to give the final answer.

partition In geometry, to divide a shape into smaller shapes. For example, shapes can be partitioned into equal shares to represent *fractions*. Partitioning can also be used to find length, *area*, or *volume*.

parts-and-total diagram A diagram you can use to represent number stories that combine two or more quantities to make a total quantity.

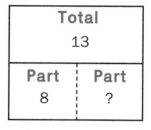

Total	
13	
Part	Part
8	?

Parts-and-total diagram for
13 = 8 + ?

pattern Shapes or numbers ordered by a rule so that what comes next can be predicted.

per "For each" or "in each." For example, "three tickets per student" means "three tickets for each student."

perimeter The distance around the boundary of a shape. The perimeter of a circle is called its circumference. The perimeter of this triangle is 15 feet.

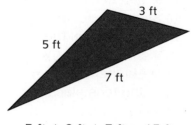

3 ft
5 ft
7 ft

5 ft + 3 ft + 7 ft = 15 ft

perpendicular Being part of two lines that cross or meet at right angles. The symbol ⊥ means "is perpendicular to."

Perpendicular lines

Perpendicular planes

Perpendicular rays

place value The value that is given to a *digit* according to its position in a number. In our *base-ten* system for writing numbers, moving a digit one place to the left makes that digit worth 10 times as much. And moving a digit one place to the right makes that digit worth one-tenth as much. For example, in the number 456, the 4 in the hundreds place is worth 400; but in the number 45.6, the 4 in the tens place is worth 40.

thousands	hundreds	tens	ones

A place-value chart

plane A flat *surface* that extends forever.

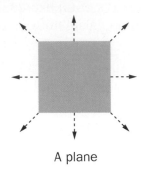

A plane

plane figure A set of points that is entirely contained in a single plane.

plot To draw on a number line or graph. The points plotted may come from *data*.

point An exact location in space. The center of a circle is a point. Lines have an unlimited number of points on them.

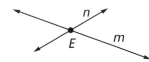

Lines *m* and *n* intersect at point *E*.

polygon A 2-dimensional figure that is made up of *line segments* joined end to end to make one closed path. The line segments of a polygon may not cross.

Polygons

polyhedron A solid whose surfaces (*faces*) are all flat and formed by *polygons*. Each face consists of a polygon and the interior of that polygon. The faces meet but may not cross. A polyhedron does not have any curved surface.

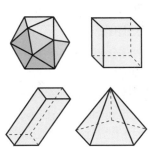

Polyhedra

positive numbers Numbers that are greater than zero. Positive numbers are usually written to the right of zero on a horizontal *number line* or above zero on a vertical number line. A positive number may be written using the + symbol, but is usually written without it. For example, +10 = 10.

population In *data* collection, the collection of people or objects that is the focus of study.

pound (lb) A U.S. customary unit equal to 16 ounces. A small can of soup weighs about 1 pound.

precise Exact. The smaller the *unit* used in measuring, the more precise the measurement is. For example, a measurement to the nearest inch is more precise than a measurement to the nearest foot. A ruler with $\frac{1}{8}$-inch markings is more precise than a ruler with $\frac{1}{4}$-inch markings.

prime number A *counting number* that has exactly two different *factors*: itself and 1. For example, 5 is a prime number because its only factors are 5 and 1. The number 1 is not a prime number because 1 has only a single factor, the number 1 itself.

prism A polyhedron with two *parallel faces*, called bases that are the same size and shape. The other faces connect the bases and are shaped like *parallelograms*. The edges that connect the bases are parallel. Prisms get their names from the shape of their bases.

Triangular Rectangular Hexagonal
prism   prism   prism

product The result of multiplying two or more numbers, called *factors*. For example, in $4 * 3 = 12$, the product is 12.

property (1) A general statement about a mathematical relationship, such as the turn-around rule or the Distributive Property of Multiplication over Addition. (2) Same as attribute.

protractor A tool on the *Geometry Template* that is used to measure and draw *angles*. The half-circle protractor can be used to measure and draw angles up to 180°; the full-circle protractor, to measure angles up to 360°.

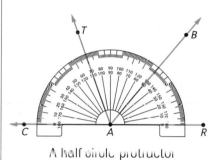

A half-circle protractor

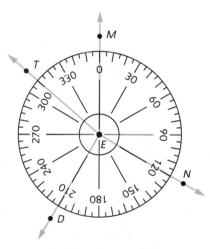

A full-circle protractor

pyramid A *polyhedron* in which one *face*, the base, may have any polygon shape. All of the other faces are triangular and come together at a point called the apex. A pyramid takes its name from the shape of its base.

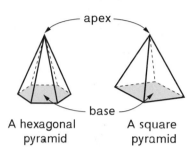

A hexagonal   A square
pyramid   pyramid

quadrangle A polygon that has four angles. Same as *quadrilateral*.

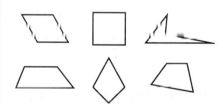

quadrilateral A *polygon* that has four sides. Same as *quadrangle*.

quantity A number with a unit, usually a measurement or count.

quart A U.S. customary unit of *volume* or *capacity* equal to 32 fluid ounces, 2 pints, or 4 cups.

quotient The result of dividing one number by another number. For example, in $35 \div 5 = 7$, the quotient is 7.

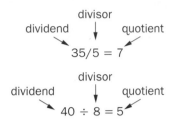

ray A straight path that starts at one endpoint and goes on forever in one direction.

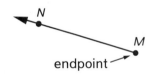

Ray *MN* or *MN*

rectangle A *parallelogram* whose corners are all *right angles*.

Rectangles

rectangular number A product of a *counting number* and the next counting number. For example, 12 is a rectangular number because $3 \times 4 = 12$.

rectangular prism A *prism* with rectangular bases. The four *faces* that are not bases are either *rectangles* or other *parallelograms*. A rectangular prism may model a shoebox.

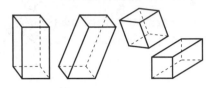

Rectangular prisms

rectilinear figure A polygon with a right angle at each vertex.

Rectilinear figures

reflection The "flipping" of a figure over a line (the line of reflection) so that its image is the mirror image of the original figure (preimage).

A reflection

reflex angle An angle with measure greater than 180° and less than 360°.

A reflex angle

regroup To rewrite a number in an equivalent form using *place-value* exchanges. For example, in a subtraction problem, 42 might need to be regrouped as $30 + 12$.

regular polygon A *polygon* whose sides are all the same length and whose interior *angles* are all the same measure. For example, a *square* is a regular polygon.

relation symbol A symbol used to express a relationship between two quantities. For example, three is less than five can be written as $3 < 5$. Some relationship symbols are $<$, $>$, and $=$.

remainder An amount left over when one number is divided by another number. For example, if 38 books are divided into 5 equal piles, there will be 7 books per pile, with 3 books left over. We may write $38 \div 5 \rightarrow 7$ R3, where R3 stands for the remainder.

represent To show, symbolize, or stand for something. For example, numbers can be represented using base-10 blocks, spoken words, or written numerals.

rhombus A *quadrilateral* whose sides are all the same length. All rhombuses are *parallelograms* and *kites*. Every square is a rhombus, but not all rhombuses are squares.

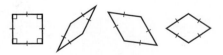

Rhombuses

right angle An *angle* with a measure of 90°.

Right angles

right triangle A triangle that has a right angle (90°).

Right triangle

rotation A movement of a figure around a fixed point; a "turn."

C

A rotation

round To change a number slightly to make it easier to work with or to make it better reflect the level of precision of the data. Often numbers are rounded to the nearest multiple of 10, 100, 1,000, and so on. For example, 12,964 rounded to the nearest thousand is 13,000.

row A horizontal ("side to side") arrangement of object in an *array*.

row

rubric A tool used to rate work based on its quality.

S

scale (1) A comparison between the number of units in a picture or model and the actual number of units. A picture graph may show 1 smiley face to stand for 10 people. (2) See *scale of a number line* or *measurement scale*. (3) A tool for measuring *weight* or *mass*.

scale of a number line The spacing of the marks on a number line. The scale of the number line below is halves.

$$\frac{0}{2} \quad \frac{1}{2} \quad \frac{2}{2} \quad \frac{3}{2} \quad \frac{4}{2} \quad \frac{5}{2} \quad \frac{6}{2}$$

scalene triangle A *triangle* with *sides* of three different lengths.

Scalene triangle

second (s or sec) (1) The basic *unit* of time. Minutes, hours, and days are based on seconds. (2) An ordinal number in the sequence first, second, third, . . .

sequence A list of numbers, often created by a rule that can be used to extend the list. Frames-and-Arrows diagrams can represent sequences.

side (1) One of the *line segments* of a *polygon*. (2) One of the rays or segments that make up an *angle*. (3) One of the *faces* of a solid figure.

situation diagram In *Everyday Mathematics*, a diagram used to organize information in a problem situation.

Total	
7	
Part	**Part**
2	5

Suzie has 2 pink balloons and 5 yellow balloons. She has 7 balloons in all.

solid See geometric solid.

solution of an open sentence A value that makes an open sentence true when it is substituted for the *variable*. For example, 7 is a solution of $5 + n = 12$.

square A *rectangle* whose sides are all the same length. A rectangle that is also a *rhombus*.

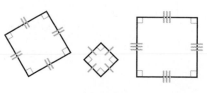

Squares

three hundred fifteen

square array A rectangular *array* with the same number of rows as columns. For example, 16 objects will form a square array with 4 objects in each row and 4 objects in each column.

A square array

square number A number that is the *product* of a counting number with itself. For example, 25 is a square number because 25 = 5 * 5. The square numbers are 1, 4, 9, 16, 25, and so on. A square number can be represented by a *square array*.

square of a number *n* The product of a number with itself. For example, 81 is the square of 9 because 81 = 9 * 9. And 0.64 is the square of 0.8 because 0.64 = 0.8 * 0.8.

square unit A *unit* used in measuring *area*, such as a square centimeter or a square foot.

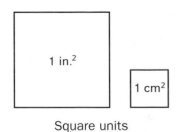

Square units

standard form The most familiar way of representing numbers. In standard form, numbers are written using the *base-ten place-value* system. For example, standard form for three hundred fifty-six is 356. Compare *expanded form*.

standard notation Same as *standard form*.

standard unit Measurement *units* that are the same size no matter who uses them and when or where they are used.

straight angle An angle that measures 180°.

A straight angle

strategy An approach to a problem that may be general, like "trial and error," or more specific, like "break apart."

sum The result of adding two or more numbers. For example, in 5 + 3 = 8, the sum is 8.

supplementary angles Two angles whose measures total 180°.

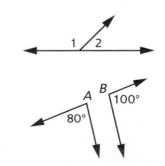

∠1 and ∠2; ∠A and ∠B are two pairs of supplementary angles.

surface (1) The boundary of a solid. Common surfaces include the top of a body of water, the outermost part of a ball, and the topmost layer of ground that covers Earth. (2) Any *2-dimensional* layer, such as a filled-in polygon.

symmetric figure (1) Having two parts that are mirror images of each other. (2) Looking the same when turned by some amount less than 360°.

symmetry A figure has line symmetry if a line can be drawn through it so that it is divided into two parts that are mirror images of each other. The two parts look alike but face in opposite directions. More generally, a figure has symmetry if parts of it look alike but are in different positions.

A figure with line symmetry

A figure with rotation symmetry

temperature A measure of how hot or cold something is.

tool Anything that can be used for performing a task. Calculators, rulers, fraction circle pieces, and number grids are examples of mathematical tools.

trade-first subtraction A subtraction method in which all trade are done before any subtractions are carried out.

trapezoid A *quadrilateral* that has at least one pair of *parallel* sides.

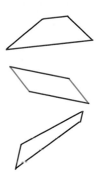

Trapezoids with parallel bases marked in the same color

triangle A *polygon* that has 3 sides and 3 angles.

Triangles

triangular number A counting number that can be shown by a triangular arrangement of dots. The triangular numbers are 1, 3, 6, 10, 15, 21, 28, 36, 45, and so on.

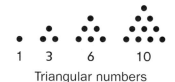

1   3   6   10

Triangular numbers

triangular prism A *prism* whose bases are triangles.

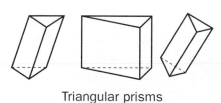

Triangular prisms

true number sentence A *number sentence* in which the relation symbol accurately connects the two sides. For example, $15 = 5 + 10$ and $25 > 20 + 3$ are both true number sentences.

turn-around rule A rule for solving addition and multiplication problems based on the Commutative Property. For example, if you know that $6 * 8 = 48$, then, by the turn-around rule, you also know that $8 * 6 = 48$.

U.S. customary system The measuring system most frequently used in the United States. See the Tables of Measures on page 292.

U.S. traditional addition algorithm An addition method that involves adding digits by place-value columns starting at the right.

$$
\begin{array}{r}
\overset{1}{}\overset{1}{} \\
3\ 4\ 8 \\
+\ \ 2\ 6\ 3 \\
\hline
6\ 1\ 1
\end{array}
$$

U.S. traditional subtraction algorithm A subtraction method that involves subtracting digits by place-value columns starting from the right, making 10-for-1 trades as needed.

$$
\begin{array}{r}
\overset{3}{}\ \overset{14}{} \\
\cancel{4}\ \cancel{4}\ 7 \\
-\ \ 1\ 6\ 5 \\
\hline
2\ 8\ 2
\end{array}
$$

unit A label used to put a number in context. In measuring length, for example, the inch and the centimeter are units. In a problem about 5 apples, *apple* is the unit.

unit conversion A change from one measurement *unit* to another using a fixed relationship, such as 1 yard = 3 feet or 1 inch = 2.54 centimeters.

unit fraction A *fraction* whose *numerator* is 1. For example, $\frac{1}{2}$, $\frac{1}{3}$, $\frac{1}{8}$, and $\frac{1}{20}$ are unit fractions.

unit square A *square* with side lengths of 1.

unknown A quantity whose value is not known. An unknown is sometimes represented by a ____, a ?, or a letter.

unlike Unequal or not the same.

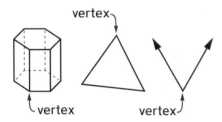

V

variable (1) A letter or other symbol that can be replaced by any number. In the number sentence $5 + n = 9$, any number may be substituted for n, but only 4 makes the sentence true. See *unknown*. (2) A number or data set that can have many values is variable.

vertex The point where the *sides* of an *angle*, the sides of a polygon, or the *edges* of a *polyhedron* meet. Plural is vertexes or vertices.

vertex

vertex

vertex

volume A measure of how much space a solid object takes up. Volume is often measured in liquid units, such as liters, or cubic units, such as cubic centimeters or cubic inches. The volume or capacity of a container is a measure of how much the container will hold.

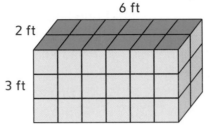

6 ft

2 ft

3 ft

36 cubic feet

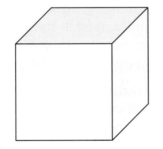

1 cubic inch

1 cubic centimeter

If the cubic centimeter were hollow, it would hold exactly 1 milliliter $\left(\frac{1}{1000}\text{ liter}\right)$.

W

weight A measure of how heavy something is.

"What's My Rule?" A type of problem with "in" numbers, "out" numbers, and a rule that changes the in numbers to the out numbers. Sometimes you have to find the rule. Other times, you use the rule to figure out the in or out numbers.

whole An entire object, collection of objects, or quantity being considered.

whole numbers The *counting numbers*, together with 0. The set of whole numbers is {0, 1, 2, 3,...}

Y

yard A U.S. customary unit of length equal to 3 feet, or 36 inches.

Page 9

Sample answer: Ethan's representation and this one use areas to represent 6 times 32. They both split 32 into 30 + 2. Ethan multiplied 6 * 30 and 6 * 2 and added the two partial products to get 192. This one splits 6 into 5 + 1 and adds four partial products: (5 * 30) + (1 * 30) + (5 * 2) + (1 * 2) to get 192.

Page 25

1. Sample answers: Using Tony's rule: 50 + 50 + 50 + 50 + 1 = 201. Using Jasmine's rule: (4 * 50) + 1 = 201.
2. The figure number is 100. I worked backwards: 401 − 1 = 400; 400 ÷ 4 = 100.
3. Subtract one from the total number of dots, then divide by 4 to find the figure number.

Page 30

1. Answers vary.
2. Answers vary.
3. Answers vary.

Page 33

1. 100 = 55 + 45
2. $\frac{1}{2} = 0.5$
3. $\frac{3}{4} > \frac{1}{4}$
4. 3 * 50 < 200
5. $\frac{1}{2} * 100 = 50$
6. 4.3 > 4.25

Page 35

1. $x = 12$
2. $z = 350$
3. 30 − (12 + 5) = 13
4. 60 = (4 + 6) * 6

Page 38

3 liters of blood

Page 44

1. Answers vary.
2. 48; Strategies vary.
3. 63; Strategies vary.

Page 46

1. 7 * 4 = 28; 2,800
2. 36 − 9 * 4; 360
3. 36 ÷ 4 = 9 or 4 * 9 = 36; 90
4. 30 ÷ 6 = 5 or 5 * 6 = 30; 50

Page 55

1. Sample answers: 12, 24, 36, 48, …
2. Sample answers: 20, 40, 60, …
3. Sample answers: 45, 90, …

Page 57

1. Sample answer: 32 = 8 * 4
2. Sample answer: 7 * 5 = 35
3. Sample answer: 5 * 3 = *t*; She planted 15 tomato plants this year.

Page 61

1. ... 60, 55, 51, 48, 46; Subtract one less from each number starting with −10
2. ... 21, 28, 36, 45, 55; Sample answers: Add 1, add 2, add 3, add 4, and so on. Starting at 1, add 2, then add the next largest counting number.
3. even
4. odd

Page 64

1.

2. Sample answer:

Page 68

1. Sample answer:

in	out
1	4
2	5
3	6
10	13
15	18

2. Sample answer:

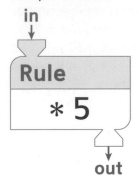

in	out
1	5
2	10
3	15
7	35
9	45

3.

in	out
9	3
36	12
1	$\frac{1}{3}$
210	70
123	41
390	130

4.

in	out
1	5
3	15
4	20
7	35
8	40
11	55

5. Sample answer:

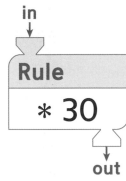

in	out
20	600

6. Sample answers: Multiplying counting numbers by 5 gives products that end in either 0 or 5. When an odd number is multiplied by 5, the product ends in a 5; when an even number is multiplied by 5, the product ends in a 0.

Page 79

1. thirty-five thousand, one hundred four; 5,000
2. seventy-one thousand, five hundred four; 500
3. three million, six hundred fifty-seven thousand; 50,000
4. eighty-two million, five hundred thousand; 500,000

Page 80

1. 203,762   2. 5,289,341
3. Sample answer: 8 [1,000s] + 7 [100s] + 4 [10s] + 4 [1s]
4. Sample answer: 1,000,000 + 400,000 + 50,000 + 6,000 + 900

Page 81

1. false   2. true   3. false

Page 87

1. 123,000; 123,500
2. 13,000; 12,700
3. 3,000; 2,900

Page 89

Sample answers:

1. 640 − 300 = 340
2. 300 + 250 + 800 = 1,350
3. 80 * 50 = 4,000
4. 140 / 7 = 20
5. Ethan has about 2,300 cards in his collection. 1,400 + 300 + 600 = 2,300

Page 91

1. Sample estimate: 330 + 250 = 580; exact sum: 579
2. Sample estimate: 70 + 45 = 115; exact sum: 112
3. Sample estimate: 300 + 100 = 400; exact sum: 421
4. Sample estimate: 2,300 + 575 = 2,875; exact sum: 2,843

5. Sample estimate: 30 + 50 + 50 = 130; exact sum: 135

Page 93

1. Sample estimate: 1,100 + 2,500 = 3,600; exact sum: 3,591
2. Sample estimate: 9,650 + 800 = 10,450; exact sum: 10,452
3. Sample estimate: 100,000 + 71,000 = 171,000; exact sum: 170,603

Page 94

1. Sample estimate: 80 − 40 = 40; exact difference: 46
2. Sample estimate: 800 − 300 = 500; exact difference: 483
3. Sample estimate: 580 − 310 = 270; exact difference: 277
4. Sample estimate: 3,600 − 1,100 = 2,500; exact difference: 2,482

Page 99

1. Sample estimate: 170 − 70 = 100; exact difference: 98
2. Sample estimate: 400 − 150 = 250; exact difference: 246
3. Sample estimate: 470 − 250 = 220; exact difference: 215
4. Sample estimate: 900 − 50 = 850; exact difference: 859

Page 101

1. Sample estimate: 440 − 380 = 60; exact difference: 55
2. Sample estimate: 310 − 280 = 30; exact difference: 30
3. Sample estimate: 8,700 − 2,700 = 6,000; exact difference: 5,986

Page 102

1. 800   2. 41,000   3. 3,000
4. 36,000   5. 2,400   6. 25,000

Page 107

1. Sample estimate: 70 * 5 = 350;
   exact product: 365
2. Sample estimate: 40 * 70 = 2,800;
   exact product: 2,709
3. Sample estimate: 40 * 30 = 1,200;
   exact product: 1,080
4. Sample estimate: 20 * 25 = 500;
   exact product: 484
5. Sample estimate: 300 * 3 = 900;
   exact product: 948

Page 110

1. 40
2. 47
3. 4,000
4. 480
5. 60

Page 114

1. Sample estimate: 390 / 3 = 130;
   exact quotient: 128
2. Sample estimate: 7)420;
   exact quotient: 63
3. Sample estimate: 2,100 ÷ 7 = 300;
   exact quotient: 263

Page 131

1. Sample answer:  ; $1\frac{2}{6}$

2. Sample answer:  ; $\frac{11}{8}$

3. Sample answer:  ; $2\frac{2}{4}$

4. Sample answer:  ; $\frac{7}{3}$

Page 135

1.
   0    $\frac{1}{2}$ 1

2.
   0         $\frac{3}{4}$ 1

3.
   0   1 $\frac{4}{3}$ 2

4. $1\frac{1}{4}$ or $\frac{5}{4}$

Page 140

A: $\frac{3}{16}$;   B: $\frac{7}{8}$, or $\frac{14}{16}$;
C: $4\frac{1}{2}$, $4\frac{2}{4}$, $4\frac{4}{8}$, or $4\frac{8}{16}$

Page 141

1. a. $\frac{2}{3}$
   b. Sample answers: $\frac{4}{6}$, $\frac{6}{9}$, $\frac{8}{12}$
2. Sample answers: $\frac{2}{8}$, $\frac{3}{12}$, $\frac{25}{100}$
3. Sample answers: $\frac{6}{10}$, $\frac{15}{25}$, $\frac{30}{50}$

Page 142

1. Sample answers: $\frac{2}{10}$, $\frac{3}{15}$, $\frac{7}{35}$
2. Sample answers: $\frac{11}{55}$, $\frac{20}{100}$, $\frac{100}{500}$

Page 144

1. $2\frac{3}{4}$ 2. $2\frac{4}{8}$, or $2\frac{1}{2}$
3. $\frac{29}{8}$ 4. $\frac{220}{100}$

Page 147

1. >   2. >   3. <   4. <

Page 151

1. 0.08, eight hundredths
2. 0.9, nine tenths
3. twenty-four and sixty-eight hundredths
4. four and six hundredths

Page 153

1. a. 200   b. 0.2 or $\frac{2}{10}$
   c. 0.02 or $\frac{2}{100}$ d. 2

2. The value of the 8 is 80. The value of the 7 is 7. The value of the 6 is 0.6 or $\frac{6}{10}$. The value of the 5 is 0.05 or $\frac{5}{100}$.

Page 155

1. >   **2.** <   **3.** <   **4.** =   **5.** >

Page 157

1. Sample answer: We cannot know, because we don't know the whole amount of money that each girl got on her birthday.

2. Sample answers: $\frac{2}{10}$, or $\frac{1}{5}$ of a lb

3. Sample answer: $\frac{2}{3}$ pizza

Page 159

1. Greater than 1. Sample answer: I started with $\frac{1}{2}$. Then I thought about how $\frac{3}{4}$ is greater than $\frac{1}{2}$. If you add more than $\frac{1}{2}$ to $\frac{1}{2}$, the total has to be greater than 1.

2. Less than 1. Sample answer: $1\frac{5}{8}$ is close to $1\frac{1}{2}$, and $\frac{9}{10}$ is close to 1. If you think of starting around $1\frac{1}{2}$ on a number line and traveling back almost a whole, you end up at a distance less than 1.

Page 161

1. $\frac{3}{4}$

2. $\frac{4}{8}$; Sample answer: When I think about $\frac{6}{8}$ on a number line, I see the point a couple eighths past the halfway mark. Then I hop back to subtract $\frac{2}{8}$.

3. $\frac{4}{4}$, or 1   **4.** $\frac{9}{10}$

Page 163

1. $6\frac{4}{5}$; Sample estimate:
$4 + 2\frac{1}{2} = 6\frac{1}{2}$

2. $3\frac{10}{8}$, or $4\frac{2}{8}$; Sample estimate:
$3 + 1\frac{1}{2} = 4\frac{1}{2}$

Page 169

1. $\frac{11}{100}$ **2.** $\frac{65}{100}$ **3.** $\frac{80}{100}$

4. $7\frac{32}{100}$ **5.** 0.47

Page 170

$5

Page 172

1. $\frac{1}{6} + \frac{1}{6} + \frac{1}{6} + \frac{1}{6} + \frac{1}{6} = \frac{5}{6}$; $5 * \frac{1}{6} = \frac{5}{6}$

2. Sample answer:

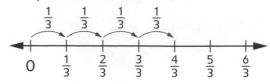

Page 174

1. $\frac{12}{5}$, or $2\frac{2}{5}$ **2.** $\frac{6}{4}$, or $1\frac{2}{4}$ (or $1\frac{1}{2}$)

3. $\frac{20}{6}$, or $3\frac{2}{6}$ **4.** $\frac{15}{8}$, or $1\frac{7}{8}$

Page 176

1. 10   **2.** $13\frac{3}{5}$ or $\frac{68}{5}$

3. $9\frac{4}{6}$ or $\frac{58}{6}$ **4.** $16\frac{1}{2}$ or $\frac{33}{2}$

Page 180

1. 4.5 cm

Page 183

1.

meters (m)	centimeters (cm)
2 m	200 cm
3.5 m	350 cm
6 m	600 cm

kilometers (km)	meters (m)
4 km	4,000 m
1.5 km	1,500 m
6 km	6,000 m

2. meters

Page 184

1. $2\frac{1}{2}$ inches

Page 187

1.

yards (yd)	feet (ft)
2 yd	6 ft
6 yd	18 ft
10 yd	30 ft

feet (ft)	inches (in.)
3 ft	36 in.
$4\frac{1}{2}$ ft	54 in.
12 ft	144 in.

2. inches, feet, or yards

Page 189

1.

kilogram (kg)	gram (g)
3 kg	3,000 g
2.5 kg	2,500 g
19 kg	19,000 g

2. cat: grams or kilograms; baseball: grams

Page 192

1.

pounds (lb)	ounces (oz)
2 lb	32 oz
7 lb	112 oz
15 lb	240 oz

tons (T)	pounds (lb)
5 T	10,000 lb
8 T	16,000 lb
20 T	40,000 lb

2. sheet of paper: ounces; a blue whale: tons; a watermelon: pounds

Page 194

1.

liters (L)	milliliters (mL)
2	2,000 mL
7	7,000 mL
39	39,000 mL

2. water in a swimming pool: liters; soup in a bowl: milliliters; water in a kitchen sink: liters

Page 197

1.

pints (pt)	cups (c)
3 pt	6 c
5 pt	10 c
10 pt	20 c

quarts (qt)	pints (pt)
1 qt	2 pt
$4\frac{1}{2}$ qt	9 pt
12 qt	24 pt

gallons (gal)	quarts (qt)
2 gal	8 qt
4 gal	16 qt
25 gal	100 qt

2. ice cream in a carton: pints, quarts, or gallons; soup in a bowl: cups or pints; water in a fish tank: gallons

Page 199

minutes (min)	seconds (sec)
1 min	240 sec
9 min	540 sec
13 min	780 sec

hours (hr)	minutes (min)
3 hr	180 min
$7\frac{1}{2}$ hr	450 min
20 hr	1,200 min

Page 201

1. 64 mm   **2.** 37 m

3. Answers vary.

Page 203

1. 6 cm²   **2.** 18 ft²

3. 48 square units

Page 206

1. 14 m²

2. 36 in.²

3. Drawings vary. Yes, the areas are the same.

4. 20 cm²

Page 209

1. 70°   **2.** 270°

Page 210

1. 25°   **2.** 150°

3.

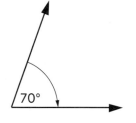
70°

4.

160°

5.

30°

Page 212

1. ∠LTU = 52°

2. ∠WXZ = 128°

3. ∠DBA = 45°

Page 214

1. 20 students   **2.** 5 students

3. none   **4.** 5 students

Page 215

1. a. smallest to largest, in inches:
$3\frac{1}{2}$, $3\frac{3}{4}$, $3\frac{3}{4}$, $3\frac{3}{4}$, $3\frac{3}{4}$, 4, 4, 4, $4\frac{1}{2}$, $4\frac{1}{2}$;
smallest: $3\frac{1}{2}$ inches; largest:
$4\frac{1}{2}$ inches

b.
Sunflower Heights
After 2 Weeks

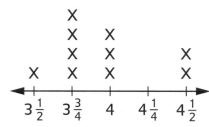

Length (inches)

c. $\frac{1}{4}$-inch intervals

d. $4\frac{1}{4}$ inches

Page 227

1.–3. Sample answer:

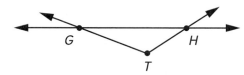

Page 229

1. Sample answer:

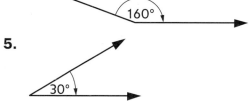

2.

3. a. ∠A   **b.** ∠C and ∠E

c. ∠B and ∠G **d.** ∠D   **e.** ∠F

f. ∠DEF, ∠FED, ∠GED, or ∠DEG

Page 230

1.–2.

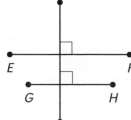

Page 231

1.
M

2.

3.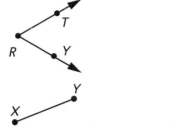

4.

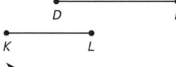

5.
D   E

K   L

6.
G
F

Page 235

1. All sides of a square have the same length. The sides of a rectangle may or may not all have the same length.

2. A rhombus and a square are both a trapezoid and a kite because both shapes have at least one pair of parallel lines AND each has two non-overlapping, adjacent pairs of equal-length sides.

Page 298

1. 30, 38, 46, 54, 62

2. 130, 114, 98, 82, 66

3. $1, 1\frac{2}{3}, 2\frac{1}{3}, 3\frac{2}{3}$

Page 299

1. $1\frac{2}{6}$ and $\frac{8}{6}$

2. $2\frac{8}{24}$ and $\frac{56}{24}$

3. $2\frac{2}{3}$ and $\frac{8}{3}$

4. $5\frac{4}{8}$ and $\frac{44}{8}$

Index

A

Abacus, 117
Abbreviations
  for area, 204
  in metric system, 179, 181, 188, 193
  for perimeter, 200, 201
  for time, 198, 292
  in U.S. customary system, 179, 185, 190,
      195, 196
Accuracy, 16–19
Acute angles, 208, 229
Adding-a-group strategy, 42, 43
Addition
  of angles, 211
  column method, 91
  of decimals, 169
  estimation in
    close-but-easier numbers, 88
    column addition, 91
    front-end method, 84
    partial-sums addition, 90
    problem solving, 83
    U.S. traditional addition, 92, 93
  of fractions
    with like denominators, 160, 169, 260
    representing, 158, 159
    with tenths/hundredths, 166–167, 169
  methods for, 90–93
  of mixed numbers, 162–163
  partial sums method, 90
  practicing, 275
  repeated addition, 8, 174, 175
  U.S. traditional, 92–93
  of whole numbers, 83–84, 88, 90–93
Addition Top-It, 275
Additive comparison, 56, 266
Air quality monitoring station, 222
Algebra, 32
Algebraic reasoning, 58
Analog clock, 198
Analytical scale, 218
Angle Add-Up, 248
Angle Race, 249
Angles
  acute, 208, 229
  adding measures of, 211
  classifying, 229
  complementary, 211
  consecutive, 235
  decomposing, 211
  defined, 228
  drawing, 208, 210, 248
  endpoints in, 208
  markings for, 234

  measuring
    classifying angles, 207, 208
    estimation in, 208, 250
    finding unknown measures,
      211–212, 248
    with a protractor, 209–210, 228
    recognizing measures of, 249
    tools for, 208, 209, 210
    units for, 207, 228
  naming, 228, 229
  non-reflex, 208
  notation for, 228, 229
  obtuse, 208, 229
  parts of, 207, 228
  in polygons, 233, 234, 235, 238
  practice with, 248, 249, 250
  reflex, 208, 229
  right, 205, 207, 208, 229
  standard unit for, 207
  straight, 211, 229
  supplementary, 211
Angle Tangle, 250
Architecture, in mathematics, 239–244
Area
  comparing with perimeter, 202
  defined, 202
  equations for, 104
  formula for, 38, 204, 293
  number sentences for, 9
  of polygons, 202
  of rectangles
    calculating, 204–206
    formula for, 204, 205, 293
    practice with, 272
    using composite units, 203
    using representations, 8–9
    using the Distributive Property, 41
  of rectilinear figures, 205–206
  of squares, 293
  units for, 202
  using composite units, 203
Area models
  with break-apart strategy, 44
  for division, 112
  doubling and, 43
  for multiplication, 103–105
Arguments, 10–11
Arrays
  adding-a-group strategy, 42, 43
  area models as, 103
  defined, 53
  for division, 111
  even numbers and, 61
  finding factors with, 53
  modeling division with, 255

multiplication squares, 43
parts of, 53
prime numbers, 54
subtracting-a-group strategy, 42, 43
turn-around rule for multiplication and, 39, 53
Arrows, 62–64
Associative Property of Multiplication, 38, 40
Atmospheric pressure, 220, 221

B

Bar codes, 76
Bar graphs, 213
Barometer, 220
Barometric pressure, 220–221
Base-10 blocks
making trades with, 153
modeling with
addition, 90, 166, 167
decimals, 154, 155
subtraction, 95
Base-10 (place value) system
decimals and, 149, 152, 153
metric system and, 179
Basic facts
division, 46, 109
multiplication, 45, 109, 271
Bath scale, 190
Beaker, 193
Beat the Calculator, 251
Benchmark, 148
Birds (real-world data), 280
Bizz-Buzz, 252
Body temperature, 222
Break-apart strategy
for multiplication, 38, 44
for renaming mixed numbers, 144
Buzz, 252

C

Calculators
electronic calculators, 118
invention of, 118
mechanical calculators, 118, 119
order of operations on, 36
practice with, 251, 254, 258, 259, 262
Capacity, 179, 193, 292
Carrying, in addition, 92
Categories, as mathematical structure, 20
Centimeter (cm), 178, 180, 202, 292
Century, 198, 292
Change diagram, 37, 51
Charts and tables. *See Tables*
Circles, degrees in, 207, 208
Cities (real-world data), 278, 282–283

Clocks, 198
Close-but-easier numbers, 88–89, 97
Collections, fractions of, 128, 156, 170
Column addition, 91
Columns, 53
Commas, in numbers, 79
Common denominators, 166, 168
Common multiples, 55
Commutative Property of Multiplication, 38, 39
Comparing and ordering
area versus perimeter, 202
data, 216
decimals, 154–155
fractions, 14–15, 145–148, 265
numbers, 81
Comparison
additive, 56, 266
multiplicative, 56–57, 266
Comparison diagram, 50, 52
Comparison number stories, 57, 266
Complementary angles, 211
Composite numbers, 54
Composite units, 203
Computers, 118, 119, 120–121
Conjectures, 10–11
Consecutive angles, 235
Conversions
for length
customary units, 17, 185, 186–187, 292
metric units, 181, 182–183, 292
for liquid volume, 173, 174, 176–177, 292
for mass, 188–189, 292
for time, 198–199, 292
for weight, 191–192, 292
Counting numbers
defined, 53, 54, 77
even/odd, 54, 61
factors of, 53
types of, 54
Counting-up subtraction, 96–99
Cubic centimeter, 292
Cubit, 178
Cup (c), 179, 195, 292

D

Data. *See also* Real-world data
bar graphs for, 213
comparing, 216
defined, 213
line plots for, 214
making sense of, 216
scaled line plots for, 215
tally charts for, 213
Day (d), 198, 292
Decade, 198, 292

Decagons, 232
Decimal point, 150, 151
Decimals
  adding, 169
  basics of, 150
  comparing, 154–155
  fractions as, 150, 151, 262
  function of, 77
  introduction to, 149
  modeling, 154, 155
  money applications, 77, 149
  naming, 150
  notation for, 149
  place value for, 152–153, 253
  practice with, 253, 263
  reading, 151
  renaming, 151, 169
  uses of, 149
  writing, 150, 153
  zero in, 155
Decimal Top-It, 253
Decimeter (dm), 292
Decomposing angles, 211
Decomposing figures, 205
Decomposing numbers, 104
Degrees, 207, 228, 229
Denominator
  common denominators, 166, 168
  defined, 125
  as divisor in division, 116
  like denominators
    adding fractions with, 160, 169, 260
    adding mixed numbers with, 162–163
    comparing fractions, 146
    subtracting fractions with, 161
    subtracting mixed numbers with,
      159, 164–165
  partitioning number lines with, 135
  patterns with, 141
  size of, 126, 127, 146
  tenths/hundredths as, 166–169
  unlike denominators, 146
Diagrams
  Frames-and-Arrows diagrams, 62–64
  multiplication/division diagrams, 48–49, 51
  situation diagrams, 50–51
Digital clock, 198
Digits, 32, 78
Dissolved oxygen in water, 219
Distance
  on a number line, 128, 133, 134
  signage for, 129
Distributive Property (of Multiplication over
    Addition/Subtraction), 38, 40–41, 44
Divide and Conquer, 254

Dividend, 111, 112, 256
Division
  area models for, 112
  basic facts, 46, 109
  estimation in, 90, 113
  extended facts, 46, 110, 254
  Fact Triangles, 254
  fractions as, 124, 128
  methods for, 111–116
  modeling, 255
  notation for, 111
  partial-quotients method, 113–114
  practicing, 254, 256, 276
  remainders in, 111, 115–116, 255
  of whole numbers, 111–116
Division Arrays, 255
Division Dash, 256
Division Top-It, 276
Divisor, 111, 112, 256
Dodecagons, 232
Doubling strategy, 43

E

Electronic computers, 119
Electronic devices, 122
Endangered species, 219
Endpoints
  in angles, 208
  of line segments, 226, 231
  on number lines, 133, 135
  of rays, 227, 231
ENIAC (Electrical Numerical Integrator and
    Calculator), 119
Equal groups
  division, 109
  equivalent fractions and, 136
  multiplication, 8–9, 39
  multiplication/division diagram for,
    48–49, 51
Equal parts
  adding fractions, 160
  comparing fractions, 147
  as decimals, 150
  fractions as, 124, 125, 128
  identifying, 126
  on number lines, 135
  unit fractions, 132
  of the whole, 125, 130, 143
Equal shares/equal sharing, 48, 109, 115, 157
Equal sign (=), 33, 81
Equal to 1, fractions as, 127
Equations
  for adding fractions, 172
  for area, 104
  defined, 33

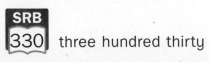

for multiplicative comparison, 56–57
for multiplying fractions, 172
Equilateral triangles, 233, 238
Equivalent Fraction Rule, 141, 148
Equivalent fractions
  defined, 136
  equal groups and, 136
  with fraction circles, 137
  identifying, 136–137, 141, 262, 263
  multiplication rule for, 141, 148
  on a number line, 136, 137, 139
  on a ruler, 139–140
  table of, 142, 293
  writing decimals for, 151
Equivalent Fractions Poster, 137
Equivalent names for numbers, 33, 268
Estimate, 52, 82, 158
Estimation
  in addition
    close-but-easier numbers, 88
    column addition, 91
    front-end method, 84
    partial-sums addition, 90
    problem solving, 83
    U.S. traditional addition, 92, 93
  of angle measures, 208, 250
  close-but-easier numbers in, 88–89, 97
  in division, 90, 113
  with fractions, 158–159, 160, 161
  front-end, 84
  of length, 16, 17, 181, 185
  of mass, 188
  with mixed numbers, 162, 163, 164, 165
  in multiplication, 84, 89, 106, 107, 108
  in problem solving, 83, 89
  purpose of, 52, 82
  rounding as, 85–87
  in subtraction
    close-but-easier numbers, 97
    counting-up method, 96–97
    modeling, 95
    with number lines, 98
    trade-first method, 94
    U.S. traditional method, 100, 101
Even numbers, 54, 61
Expanded form, 80
Extended facts
  division, 46, 110, 254
  multiplication, 45, 102, 251

F

Factor Bingo, 257
Factor Captor, 258
Factors
  in area models, 103
  breaking apart, 38, 44
  in composite numbers, 54
  of counting numbers, 53
  defined, 53
  finding, 53, 258
  identifying, 53, 257
  multiples and, 55
  number sentences with missing factors, 110
  order for multiplying, 39
  in partial-products multiplication, 106–107
  in prime numbers, 54
  products of more than two, 38, 40
Facts
  basic facts
    division, 46, 109
    multiplication, 45, 109, 271
  extended facts
    division, 46, 110, 254
    multiplication, 45, 102, 251
  helper facts
    adding-a-group strategy, 42, 43
    Associative Property and, 40
    break-apart strategy, 44
    subtracting-a-group strategy, 42, 43
  practice with, 251, 254, 271
  strategies for, 42–44
Fact Triangles, 254
Fair trades, 144
Fathom, 178
Fibonacci sequence, 58
Finger stretch, 178
Fish, protecting, 214
Fishing for Digits, 259
Fishing for Fractions, 260–261
5-gallon bucket, 195
Fluid ounce (fl oz), 292
Foods (real-world data), 286
Food supplies (real-world data), 287
Foot (ft), 179, 184, 202, 292
Formula
  for area, 38, 204, 293
  defined, 38
  for perimeter, 200, 201, 293
Fraction circles
  adding fractions with, 160
  comparing fractions with, 145, 146, 147, 148
  equivalent fractions with, 136, 137
  in estimation, 158
  introduction to, 130–131
  as mathematical tool, 15
  multiplying fractions with, 173
  problem solving with, 157
  renaming fractions with, 143
  representing fractions with, 130–131

Fraction/Decimal Concentration, 262
Fractions. *See also* Denominator; Numerator
  defined, 124
  adding
    with like denominators, 160, 169, 260
    representing, 158, 159
    with tenths/hundredths, 166–167, 169
  benchmark fractions, 148
  of collections, 128, 156, 170
  comparing, 14–15, 145–148, 265
  decimals as, 150, 262
  as division, 124, 128
  equal parts
    adding fractions, 160
    comparing fractions, 147
    fractions as, 124, 125, 128
    unit fractions, 132
    of the whole, 125, 130, 143
  equal to 1, 127
  equivalent
    defined, 136
    equal groups and, 136
    with fraction circles, 136, 137
    identifying, 136–137, 141, 262, 263
    multiplication rule for, 141, 148
    on a number line, 136, 137, 139
    on a ruler, 139–140
    table of, 142, 293
    writing decimals as, 136
  estimating with, 158–159, 160, 161
  function of, 77
  greater than 1
    defined, 127
    on number lines, 134
    representing 131, 132, 143, 144
  history of, 124
  meanings of, 128
  multiplying whole numbers by, 171–172,
    173–174, 264
  notation for, 125, 128
  on number lines
    distance on a number line, 128
    estimating on, 159
    modeling, 13, 14–15
    Number-Lines Poster, 137, 138
    plotting on, 133–135
  practice with, 260–265
  problem solving with, 156–157
  reading, 125–126
  reasoning with, 156–157
  remainders as, 116
  renaming, 143, 151, 163, 166, 167
  representing with fraction circles, 130–131
  simplest form for, 142
  subtracting with like denominators, 161

  unit fractions
    adding/subtracting fractions as, 160–161
    defined, 132, 171
    multiples of, 171–172, 174
  uses of, 124, 129
  writing, 125–126
Fraction strips, 14, 15, 132
Fraction Top-It, 265
Frames, 62–64
Frames-and-Arrows diagrams, 62–64
Frequency, in sound, 69, 70, 74
Friendly numbers, 88
Front-end estimation, 84
Full-size protractor, 208, 209, 228, 236
Function machines, 37, 65–68

G

Gallon (gal), 179, 195, 292
Games
  Addition Top-It, 275
  Angle Add-Up, 248
  Angle Race, 249
  Angle Tangle, 250
  Beat the Calculator, 251
  Bizz-Buzz, 252
  Buzz, 252
  Decimal Top-It, 253
  Divide and Conquer, 254
  Division Arrays, 255
  Division Dash, 256
  Division Top-It, 276
  Factor Bingo, 257
  Factor Captor, 258
  Fishing for Digits, 259
  Fishing for Fractions, 260–261
  Fraction/Decimal Concentration, 262
  Fraction Match, 263
  Fraction Multiplication Top-It, 264
  Fraction Top-It, 265
  How Much More?, 266
  introduction to, 246–247
  Multiplication Top-It, 276
  Multiplication Wrestling, 267
  Name That Number, 268
  Number Top-It, 269
  Polygon Capture, 270
  Product Pile-Up, 271
  Rugs and Fences, 272
  Spin-and-Round, 273
  Subtraction Target Practice, 274
  Subtraction Top-It, 275
Geometry
  angles
    acute, 208, 229

adding, 211
classifying, 208, 229
defined, 228
drawing, 208, 210, 248
markings for, 234
measuring
    classification of, 207, 208
    estimation in, 208, 250
    finding unknown measures, 211–212, 248
    with a protractor, 209–210, 228
    recognizing measures of, 249
    tools for, 208, 209, 210
    units for, 207, 228
naming, 228, 229
obtuse, 208, 229
parts of, 207, 228
in polygons, 232, 233, 234, 235, 238
reflex (non-reflex), 208, 229
right, 205, 207, 208, 229
straight, 211, 229
decagons, 232
dodecagons, 232
endpoints
    in angles, 208
    of line segments, 226, 231
    on number lines, 133, 135
    of rays, 227, 231
heptagons, 232
hexagons, 232, 235
intersecting lines, 230, 231
intersecting line segments, 200, 201
kites, 235
line of symmetry, 238
lines, 227, 231
line segments, 226, 228, 231
mathematics and architecture (photo essay),
    239–244
nonagons, 232
notation in, 226, 227, 228, 229, 230,
    231, 233
octagons, 232, 238
parallel lines, 230, 231
parallel line segments, 230, 231
parallelograms, 234, 235
patterns in, 224–225
pentagons, 232, 238
perpendicular lines, 230, 231
perpendicular line segments, 230, 231
points, 226, 231
polygons, 232
    classifying, 232
    defined, 232
    naming, 232
    parts of, 232, 233, 234, 235

perimeter of, 200
    properties of, 232, 270
pyramids, 242, 243
quadrilaterals, 232, 234
rays, 207, 208, 227, 228, 231
real-world applications, 224–225
rectangles (See Rectangles)
rhombuses, 235
squares, 201, 235, 238
symmetric figures, 238, 241, 242
trapezoids, 234, 235
triangles, 232, 233
Geometry Template, 236, 237
Goals for Mathematical Practice (GMPs), 2–3, 4,
    5, 6, 7, 9, 10, 11, 13, 15, 17, 18, 19,
    20, 21, 25
Graduated cylinder, 193, 218
Grains, 178
Gram (g), 179, 188, 292
Graphing
    bar graphs, 213
    line plots, 214
    scaled line plots, 215
    tally charts, 213
Greater than 1
    defined, 127
    on number lines, 134
    representing 131, 132, 143, 144
Greater-than symbol (>), 33, 81, 229
Great span, 178
Grouping symbols, 34–35
Guide to Solving Number Stories, 26

H

Half-circle protractor, 208, 210, 228, 236
Halfway numbers, 87
Helper facts
    adding-a-group strategy, 42, 43
    Associative Property and, 40
    break-apart strategy, 44
    subtracting-a-group strategy, 42, 43
Heptagons, 232
Hertz (Hz), 69
Hexagons, 232, 238
Higher number, in rounding, 86
Hour (hr), 198, 292
How Much More?, 266
Humidity, 220
Hundredths
    adding, 166–169
    depicting, 150
    place value of, 152, 153
    reading, 151
    renaming fractions as decimals, 151
Hundredths grid, 150, 169

Hurricanes, 221
Hygrometer, 220

 I

Identity Property of Multiplication, 38, 39
Improper fractions, 127
Inch (in.)
  as customary unit, 179, 184
  problem solving with, 17–18
  on rulers, 178
  square inch, 202
  unit conversions, 292
Indoor water parks (real-world data), 280
Inequalities, 33
In/out rule, 65–68
Interpreting remainders, 115–116
Intersecting lines, 230, 231
Intersecting line segments, 230, 231
Isosceles trapezoids, 235

 J

Joint (natural measure), 178

 K

Keyboard instruments, 74
Kilogram (kg), 178, 188, 292
Kiloliter (kL), 292
Kilometer (km), 179, 180, 292
Kites, 235

 L

Large numbers
  adding/subtracting, 275
  partitioning, 105
  place value of, 78–79
Lattice, 108
Lattice multiplication, 108, 117
Length
  ancient measures of, 178
  area formula, 38, 204, 293
  of the boundary, 202
  decimal measures of, 149
  estimating, 16, 17, 181, 185
  fractional measures of, 140
  importance of, 217
  metric system units, 180–183
  for perimeter, 200–201
  personal references in, 181, 185
  standard units for, 178, 179
  tools for measuring, 180, 184
  unit conversions in
    customary units, 17, 185, 186–187, 292
    metric units, 181, 182–183, 292
  U.S. customary system units, 184–187

Less-than symbol (<), 33, 81
Like denominators
  adding fractions, 160
  adding mixed numbers, 162–163
  comparing fractions, 146
  subtracting fractions, 161
  subtracting mixed numbers, 159, 164–165
Line of symmetry, 238. See also Symmetric
    figures
Line plots, 214, 216
Lines, 227, 231
Line segments, 205, 226, 228, 231
Liquid volume
  defined, 193
  personal references in, 193, 195
  standard units for, 179, 193, 195
  tools for measuring, 193, 195, 218
  unit conversions in, 193–194,
    196–197, 292
Liter (L), 179, 193, 292
Longest roller coasters in the world
    (real-world data), 279
Lowest terms, 142

 M

Machines that calculate, 117–122
Maps, as real-world data, 282–284
Mass
  defined, 188, 218
  estimating, 188
  personal references in, 188
  standard units for, 179, 188
  tools for measuring, 188, 218
  unit conversions in, 188–189, 292
Mathematical model, 12–13
Mathematical practices, 1–30
Mathematical reasoning, 10, 25
Mathematical representations, 8–9
Mathematical symbols, 32
Mathematical tools. See also specific tools
  machines that calculate, 117–122
  for measuring angles, 208, 209, 210
  for measuring length, 180, 184
  for measuring liquid volume, 193, 195
  for measuring weight, 190
  for problem solving, 14–15
Mathematics and architecture (photo essay),
    239–244
Maximum (data value), 215, 216
Measurement. See also Metric system;
    U.S. customary system
  of angles
    classifying angles, 207, 208
    estimation in, 250
    finding unknown measures,
      211–212, 248

with a protractor, 209–210, 228
recognizing measures of, 249
units for, 207, 228
fractional measures, 140
of length, 180–183, 184–187
natural measures, 178
in the natural world (photo essay),
217–222
standard units, 178
tools for, 180, 184, 190, 193, 195
Measurement scale
for graphing, 213, 215
for length, 182–183, 186–187
for liquid volume, 193–194, 196–197
for mass, 188–189
for time, 198–199
for weight, 191–192
Measuring cup, 129, 195
Measuring spoon, 129, 195
Meter (m), 179, 202, 292
Metric system. See also specific units
abbreviations in, 179, 181, 188, 193
decimals in, 149
introduction to, 179
length measurements, 180–183
liquid volume measurements, 193 194
mass measurements, 188–189
personal references in, 181, 188, 193
Metric ton (t), 292
Micrometer, 217
Mile (mi), 179, 184, 202, 202
Millibars (mb), 220
Milligram (mg), 218, 292
Milliliter (mL), 179, 193, 292
Millimeter (mm), 179, 180, 292
Minimum (data value), 215, 216
Minute (min), 198, 292
Mixed numbers. See also Fractions
adding, 162–163, 168, 261
decimals as, 150
decomposing, 163, 164, 165, 168
defined, 127, 143
estimating with, 162, 163, 164, 165
with fraction circles, 131
with fraction strips, 132
multiplying by whole numbers, 175–176
renaming, 144, 176
on rulers, 139, 140
subtracting, 159, 164–165, 261
writing, 127
Modeling
with area models
with break-apart strategy, 44
for division, 112
doubling, 43
for multiplication, 103–105

with base-10 blocks
modeling addition, 90, 166, 167
modeling decimals, 154, 155
modeling subtraction, 95
with counters, 170
division with arrays, 255
mathematical model, 12–13
with number lines, 13, 14–15
Money
decimal values of, 77, 149
fractional values of, 129
Monthly precipitation (real-world data), 285
Months (mo), 198, 292
Multiples
common multiple, 55
in counting-up subtraction, 95
defined, 55
extended facts and, 45–46, 102, 110
finding, 55, 252
renaming fractions, 167
skip counting with, 55
of ten, 179
unit fraction multiples, 171–172, 174
Multiplication
adding-a-group strategy, 42, 43
area models for, 103–105
Associative Property of Multiplication, 38, 40
basic facts, 45, 109, 271
break-apart strategy, 38, 44
by 1, 39
Commutative Property of Multiplication, 38, 39
diagram for, 48–49, 51
Distributive Property (of Multiplication over
Addition/Subtraction), 38, 40–41
doubling strategy, 43
equal groups, 8–9, 39
estimation in, 84, 89, 106, 107, 108
extended facts, 45, 102, 251
fact strategies, 42–44
of fractions by whole numbers, 171–172,
173–174, 264
Identity Property of Multiplication, 38, 39
lattice method, 108, 117
methods for, 106–108
of mixed numbers by whole numbers,
175–176
with more than two factors, 38, 40
multiplicative comparison, 56–57, 266
near-squares strategy, 43
number models for, 28
partial-products method
Associative Property and, 40
method of, 106–107
for multiplying mixed numbers, 176
practicing, 267
in problem solving, 9

Multiplication (*continued*)
   practicing, 264, 267, 271, 276
   properties of, 38-41
   shortcuts in, 102
   subtracting-a-group strategy, 42, 43
   turn-around rule for, 38, 39, 53
   of whole numbers
      by fractions, 171–172, 173–174, 264
      methods for, 106–108
      by mixed numbers, 175–176
   Zero Property of Multiplication, 38
Multiplication and division table, 291
Multiplication/division diagram, 48–49, 51
Multiplication rule for equivalent fractions,
      141, 148
Multiplication square, 43
Multiplication Top-It, 276
Multiplication Wrestling, 267
Multiplicative comparison, 56–57, 266
Multiplicative Identity Property. *See* Identity
      Property of Multiplication
Music, in mathematics, 59, 69–74

Name-collection boxes, 33
Name That Number, 268
Napier's bones, 117
Natural measures, 178
Natural yard, 178
Near-squares strategy, 43
Negative numbers, 77
Nonagons, 232
Non-reflex angles, 208
Normal September rainfall (real-world data), 284
Notation
   for angles, 228, 229
   for decimals, 149
   for fractions, 125, 128
   for lines, 227, 231
   for line segments, 226, 231
   for parallel lines, 230, 231
   for perpendicular lines, 230, 231
   for points, 226, 231
   for rays, 227, 231
   for triangles, 233
Not-equal symbol ($\neq$), 33, 81
Number grids, patterns on, 60
Number lines
   comparing fractions on, 145, 146, 148
   counting up on, 98–99
   equivalent fractions on, 136, 137, 139
   finding points on, 133–134
   fractions on
      distance on a number line, 128
      estimating on, 159

      modeling, 13, 14–15
      Number-Lines Poster, 137, 138
      plotting on, 133–135
   modeling with, 13, 14–15
   multiplying fractions on, 173
   partitioning of, 133–135
   plotting on, 135
   renaming fractions on, 143
   renaming mixed numbers on, 144
   rounding on, 85
   subtracting fractions on, 161
   subtracting mixed numbers on, 164
   unit fractions on, 172
   zero on, 133
Number-Line Poster, 137, 138
Number models
   in algebra, 32
   for change situations, 37, 51
   for comparison situations, 50, 52
   defined, 37, 47
   for division, 112
   for dot patterns, 24
   for equal-groups situations, 48
   for equal-shares situations, 49
   for multiplication, 28
   for multiplicative comparison, 56, 57
   for number stories, 37, 47, 48, 50
   for parts-and-total situations, 50
   for perimeter, 19
   with remainders, 49
Numbers
   close-but-easier, 88–89, 97
   comparing, 81
   composite, 54
   equivalent names for, 33, 268
   estimating, 82–84
   even/odd, 54, 61
   factors for, 54
   kinds of, 77
   multiples of
      common multiple, 55
      in counting-up subtraction, 95
      defined, 55
      extended facts and, 45–46, 102, 110
      finding, 55, 252
      skip counting with, 55
      of ten, 179
   name-collection boxes, 33
   patterns in, 20–21, 58–61
   place value of, 78–79, 259, 269
   prefixes for, 291
   prime, 54
   reading, 76, 77
   rounding, 85–87
   standard form/notation, 80

triangular, 61
uses of, 76
writing, 76, 77
Number sentences
for adding fractions, 167
for area, 9
for comparison situations, 52
defined, 32
for dot patterns, 24
evaluating, 32, 33, 35, 109
with fractions, 127
with missing factors, 110
for number patterns, 21
open sentences, 32
parentheses in, 34–35
for perimeter, 19
for repeated addition, 8, 9
for sequences, 58
solving, 32
Number stories. See also Problem solving
comparison number stories, 57, 266
diagrams for, 27, 50–51
guide to solving, 26
number models for, 37, 47, 38, 50
organizing information in, 48, 50, 51
1-step, 50
2-step, 51
Number Top-It, 269
Numerator
adding/subtracting fractions, 160–161
comparing fractions, 146, 147
defined, 125
fractional parts on number lines, 135
like numerators, 146, 147
patterns with, 141
as remainder in division, 116
size of, 126, 127, 146
unlike numerators, 148

O

Obtuse angles, 208, 229
Octagons, 232, 238
Odd numbers, 54, 61
1-finger width, 178
1-step number stories, 50
Open number lines, 98–99
Open sentences, 32
Operation symbols, 32
Order of operations, 34, 35, 36, 291
Ounce (oz), 179, 188, 190, 292

P

Pan balance, 188, 218
Parallel lines, 230, 231

Parallel line segments, 230, 231
Parallelograms, 234, 235
Parentheses
with Associative Property, 40
with Distributive Property, 40
in number sentences, 34–35
Partial-products multiplication
Associative Property and, 40
method of, 106–107
for multiplying mixed numbers, 176
practicing, 267
in problem solving, 9
Partial quotient, 113
Partial-quotients division, 113–114
Partial-sums addition, 90
Partitioning
of number lines, 133–135
of rectangles, 8, 9, 104–105, 106, 112
Parts-and-total diagram, 50
Pattern-block shapes, 236
Patterns
architectural, 240, 241
dot patterns, 22–25, 61
in equivalent fractions, 141
geometric, 224–225
making predictions, 59
as mathematical structure, 20
in music, 59, 74
in numbers, 20–21, 58–61
repeating patterns, 59
in rounding, 87
in "What's My Rule?", 66, 67, 68
Pentagons, 232, 238
Per, defined, 48
Percussion instruments, 69, 72
Perimeter
compared with area, 202, 272
defined, 200
finding, 272
formulas for, 200–201, 293
problem solving with, 17–19
Perpendicular lines, 230, 231
Perpendicular line segments, 230, 231
Personal computers, 120–121
Personal references in measurement
in metric system, 181, 188, 193
in U.S. customary system, 185, 191, 195
Photo essays
machines that calculate, 117–122
mathematics and architecture, 239–244
measurements in the natural world, 217–222
sound, music, and mathematics, 69–74
Pint (pt), 179, 195, 292
Pitch, in sound, 70–71, 72, 73

Index

Place value
    for decimals, 152–153, 253
    defined, 78
    expanded form, 80
    extended facts and, 46
    of numbers, 78–79, 259, 269
Place-value chart, 78–79, 108, 152, 153,
        155, 291
Points, 226, 231
Polygon Capture, 270
Polygons
    classifying, 232
    defined, 232
    naming, 232
    perimeter of, 200, 293
    properties of, 232, 270
    regular, 238
    sides/angles of, 232, 233, 234, 235, 238
    vertices of, 232
Population (real-world data), 282
Pound (lb), 178, 179, 188, 190, 292
Precipitation (real-world data), 284, 285
Precision, 16–19, 218
Prime numbers, 54
Problem solving
    accuracy in, 16–19
    arguments in, 10–11
    conjectures in, 10–11
    diagram for, 27
    estimation in, 83, 89
    with fractions, 156–157
    Guide to Solving Number Stories, 26
    making sense of problem, 4–8, 50
    mathematical models in, 12–13
    mathematical practices for, 1–30
    persevering in, 5–7
    precision in, 16–19
    process of, 26–27
    representations in, 8–9
    shortcuts for, 22–25
    structure in, 20–21
    tools for, 14–15
    using diagrams (drawings) in, 4–6, 8, 12–14,
        16–17, 20–23, 29, 57
    using tables in, 21
Product, 39, 53, 103
Product Pile-Up, 271
Properties of operations
    Associative Property of Multiplication, 38, 40
    Commutative Property of Multiplication,
        38, 39
    defined, 39
    Distributive Property (of Multiplication over
        Addition/Subtraction), 38, 40–41
    Identity Property of Multiplication, 38, 39
    as mathematical structure, 20
    multiplication, 38, 39–41
Protractor, 208, 211, 228
Pulse rate, 222
Pyramids, 242, 243

Q

Quadrangles, 234
Quadrilaterals, 232, 234
Quart (qt), 179, 195, 292
Quotient, 111, 256

R

Rainfall (real-world data), 284–285
Rain gauge, 220
Ratios, fractions as, 124
Rays, 207, 208, 227, 228, 231
Real-world data
    data sources, 289–290
    foods, 286
    food supplies, 287
    indoor water parks, 280
    introduction to, 278
    monthly precipitation, 285
    normal September rainfall, 284
    roller coaster length, 279
    songbird wing length, 288
    U.S. city population, 278, 282–283
    zoo size, 278, 281
Rectangles
    adding fractions with, 160
    area of
        calculating, 204–206
        formula for, 204, 205, 293
        practice with, 272
        using composite units, 203
        using representations, 8–9
        using the Distributive Property, 41
    multiplying fractions with, 173
    partitioning, 8, 9, 104–105, 106, 112
    perimeter of, 17, 19, 200, 272, 293
    properties of, 235
    subtracting fractions with, 161
Rectilinear figures, 205–206
Reflections. *See* Line of symmetry
Reflex angles, 208, 229
Regular hexagon, 238
Regular octagon, 238
Regular pentagon, 238
Regular polygon, 238
Relations, 32, 33
Remainders, 111, 115–116, 255
Repeated addition, 8, 174, 175
Repeating patterns, 59

Representations
    for adding fractions, 158, 159
    creating, 8, 9
    defined, 8
    fraction circles, 130–131, 143, 144, 158
    fraction strips, 132
    making connections between, 9
    making sense of, 9
Rhombuses, 235
Right angles, 205, 207, 208, 229
Right triangles, 233
Roller coaster length (real-world data), 279
Rotation, of angles, 207
Rounding, 85–87, 273
Rows, 53
Rugs and Fences, 272
Rulers
    defined, 226
    equivalent fractions on, 139–140
    fractions on, 129
    on Geometry Template, 236
    as measuring tool, 180, 184, 217
Rules
    for Frames-and-Arrows diagrams, 62–64
    in/out, 37
    multiplication rule for equivalent fractions,
        141, 148
    "What's My Rule?", 37, 65–68

S

Satellites, weather technology, 221
Scaled bar graph, 213
Scaled line plots, 215
Science
    health and, 222
    process of, 217
    weather, 220–221
Second (sec), 198, 292
September precipitation (real-world data), 284
Sequence, 58, 60
Shapes. See specific shapes
Sides
    of angles, 228
    markings for, 234
    of polygons, 232, 233, 234, 235, 238
    of squares, 201
    of triangles, 233
Simplest form, 142
Situation diagrams, 50–51
Sizes of indoor water parks (real-world data), 280
Sizes of zoos around the world (real-world data),
        278, 281
Skip counting, 55, 109, 111
Solution, 32
Solving the number sentence, 32

Songbird wing length (real-world data), 288
Sound, in mathematics, 69–74
Spin-and-Round, 273
Spring scale, 190
Square centimeters, 202
Square feet, 202
Square inches, 202
Square meters, 202
Square miles, 202
Squares
    area of, 293
    line of symmetry for, 238
    perimeter of, 201, 293
    properties of, 235
Square units (unit²), 202, 204
Standard form, 80
Standard masses, 218
Standard notation, 80
Standards for Mathematical Practice, 1–30
Standard units, 178
Straight angle, 211, 229
Straightedge, 226
Stringed instruments, 69, 73
Structure, in mathematics, 20–21
Subtracting-a-group strategy, 42, 43
Subtraction
    counting-up method, 96–99
    Distributive Property with, 41
    estimation in
        close-but-easier numbers, 97
        counting-up method, 96, 97
        modeling, 95
        with number lines, 98
        trade-first method, 94
        U.S. traditional method, 100, 101
    of fractions with like denominators, 161, 261
    methods for, 94–101
    of mixed numbers, 159, 164–165, 261
    practicing, 261, 274, 275
    trade-first, 94
    U.S. traditional, 100–101
    of whole numbers, 94–101
Subtraction Target Practice, 274
Subtraction Top-It, 275
Supplementary angles, 211
Surface inside the boundary, 202
Symmetric figures, 238, 241, 242

T

Tables
    comparing numbers, 81
    of equivalent fractions, 142, 293
    of factors, 54, 55
    food supplies data, 287
    in/out, 65–68

Index

Tables (*continued*)
  multiplication and division, 291
  of number models, 47
  number prefixes, 291
  order of operations, 36, 291
  place-value chart, 78–79, 108, 152, 153,
     155, 291
  for problem solving, 21
  properties of multiplication, 38
  rainfall data, 285
  relation symbols, 33
  roller coaster data, 279
  songbird data, 288
  for unit conversions
    length, 182–183, 186–187, 292
    liquid volume, 193–194, 196–197, 292
    mass, 188–189, 292
    time, 198–199, 292
    weight, 191–192, 292
  water park data, 280
  "What's My Rule?", 37, 65–68
  zoo data, 281
Tablespoon (tbs), 292
Tablet computers, 121
Tally charts, 213
Tape measure, 217
Teaspoon (tsp), 292
Temperature
  of the body, 222
  decimals with, 149
  measuring, 219, 220
Tenths
  adding, 166–169
  depicting, 150
  place value of, 152, 153
  reading, 151
  renaming fractions as decimals, 151
Tenths grid, 150
Thermometer, 220, 222
The whole. *See also* Fractions
  breaking apart, 144
  comparing fractions, 145
  defined, 124
  equal parts of, 125, 130, 143
  with fraction circles, 130–131
  with fraction strips, 132
  identifying, 156
  parts of, 150
  renaming fractions, 143
  renaming mixed numbers, 144
Tiling, 204
Time, 198–199
Ton (T), 179, 190, 292
Tools, 14–15. *See also* Mathematical tools
Tornadoes, 221
Trade-first subtraction, 94

Trapezoids, 234, 235
Triangles, 232, 233
Triangular numbers, 61
Turn-around rule for multiplication
  in arrays, 39, 53
  defined, 38, 39
2-finger width, 178
2-step number stories, 51

U

Unit conversions
  length
    customary units, 17, 185,
     186–187, 292
    metric units, 181, 182–183, 292
  liquid volume, 193–194, 196–197, 292
  mass, 188–189, 292
  time, 198–199, 292
  weight, 191–192, 292
Unit fractions
  adding/subtracting fractions as, 160–161
  defined, 132, 171
  multiples of, 171–172, 174
Units of measurement. *See* Metric system;
    U.S. customary system
UNIVAC (Universal Automatic Computer), 119
Unknowns. *See* Variables
U.S. city population (real-world data), 278,
    282–283
U.S. customary system. *See also specific units*
  abbreviations in, 179, 185, 190, 195, 196
  introduction to, 179
  length conversions in, 17, 185, 186–187, 292
  length measurement in, 184–187
  liquid volume in, 195–197
  personal references in, 185, 191, 195
  weight measurements, 190–192
U.S. traditional addition, 92–93
U.S. traditional subtraction, 100–101

V

Value, of digits, 78
Variables
  defined, 32, 37
  in formulas, 38, 293
  in properties of operations, 38
  in rules, 37
  solving for, 34–35, 36
  using, 32, 37
Vertex (vertices)
  of angles, 207, 228
  defined, 232
  of polygons, 232, 233, 234
  of triangles, 233
Volume, defined, 218. *See also* Liquid volume

Water, quality of, 222
Water parks (real-world data), 280
Weather, 220–221
Week (wk), 198, 292
Weight. *See also* Mass
    ancient measures of, 178
    defined, 190
    personal references in, 191
    standard units for, 178, 179, 190
    tools for measuring, 190
    unit conversions in, 191–192, 292
What do Americans eat? (real-world data), 286
"What's My Rule?", 10–11, 37, 65–68
Whole. *See* The whole
Whole numbers
    adding, 83–84, 88, 90–93
    defined, 77
    dividing, 111–116
    fractions between, 124
    multiplying
        by fractions, 171–172, 173–174, 264
        methods for, 106–108
        by mixed numbers, 175–176
    place value for, 78–79, 259, 269
    rounding of, 85
    subtracting, 94–101
Width, 200, 204
Wind instruments, 69, 70, 71
Wind speed/direction, measuring, 220
World food supplies (real-world data), 287
World Wide Web, invention of, 121

Yard (yd), 17, 18, 179, 184, 292
Year (yr), 198, 292

Zero
    in decimals, 155
    in division, 109
    fractions and, 125
    in front-end estimation, 84
    in multiplication, 109
    on number lines, 133
    as whole number, 77
Zero Property (of Multiplication), 38
Zoos (real-world data), 278, 281